GCSE
Chemistry
for You

W E Latchem

Hutchinson

London Melbourne Sydney Auckland Johannesburg

Acknowledgements:

Acknowledgement is due to the following for permission to reproduce photographs:
C. Ashall p.245; Australian News and Information Bureau p.48; BBC Hulton Picture Library p.62; Blackwood Hodge p.9; Antony Blake p.247; J. Brown p.62; British Airways p.112; British Alcan Aluminium Ltd. pp.256, 269; British Gas p.129; British Petroleum pp.8, 28, 29, 97, 107; British Steel pp.56, 81, 256, 274, 275; British Sugar pp.9, 27; British Sulphur Corporation Ltd. p.136; British Tourist Authority pp.18, 81, 100; Boots Company PLC p.178; Camera Press pp.63, 87, 113; J. Allan Cash pp.185, 244; Copper Development Association pp.68, 69, 73, 156; Crompton Parkinson Ltd. p.149; Daily Telegraph p.56; De Beers pp.126, 127; The Distiller's Co. PLC p.208; European Ferries Ltd. p.6; FAO pp.179, 245; Farmer's Weekly p.48; Fison's Pest Control Ltd. p.163; The Flour Advisory Bureau p.209; Grant's of St. James's pp. 206, 208; Griffin and George pp.15, 26, 60; The Guardian p.148; Phillip Harris Ltd. p.403; Hawker Siddeley p.73; ICI Agricultural Division pp.234, 245 ICI Corporate Slide Bank pp.15, 46, 146, 172, 179, 219, 220, 221, 233, 235, 236, 237;

Imperial War Museum p.21; Industrial Diamond Review p.202; Institute of Dermatology p.178; Institute of Geological Sciences p.202; John Lewis Partnership p.74; Lead Development Association p.73; Mansell Collection p.103, May and Baker Ltd. p.19; Metropolitan Police p.209; John Moss p.18; NASA p.7, National Coal Board pp.8, 48, 128, 129; National Geological Survey p.8; National Water Council p.102; Nuffield-Chelsea Curriculum Trust p.134; Nuffield Foundation (from Revised Chemistry Handbook for Pupils, Longman 1978) pp.131, 134; Nu-Swift International Ltd. p.91; GFC Overton pp.75, 86, 142, 215; Mary Quant p.217; Photo Source p.13; Pilkington Brothers Ltd. p.87; Portsmouth and Sunderland Newspapers Ltd. pp.50, 62, 81, 124, 148; Portsmouth Water Company p.102; R.H.M. Foods Ltd. p.108; Robson and Baxter Ltd. p.14; Lisa Taylor/Save the Children p.247; Schweppes Ltd. p.91; Science Museum pp.38; Space Frontiers Ltd. p.185; St. Mary's Hospital Medical School p.178; Thames Water Authority pp.92, 103; The Times p.105; Tungstone Batteries Ltd. p.149; Unilever Research Laboratory p.7; Wella p.217; West Air Photography p.7; C.B. Wilberforce p.62.

The author also wishes to thank M.J. Latchem for providing artwork throughout the book.

This book is made up of units, usually of two pages, which deal with one topic. Several related topics make up a section. Experiments are numbered with the section number.

Important Note
Experiments marked * should only be done by a teacher.

Hutchinson Education

An imprint of Century Hutchinson Ltd.
62 – 65 Chandos Place, London WC2N 4NW

Century Hutchinson Group (Australia) Pty Ltd.
16 – 22 Church Street, Hawthorn, Melbourne, Victoria 3122

Century Hutchinson Group (NZ) Ltd.
32 – 34 View Road, P.O. Box 40–086, Glenfield, Auckland 10

Century Hutchinson Group (SA) (Pty) Ltd.
P.O. Box 337, Bergvlei 2012, South Africa

First published as Chemistry For You Books One (1982) and Two (1983)

GCSE Edition first published 1986

© W E Latchem 1982, 1983, 1986

Designed and typeset by The Pen & Ink Book Company Ltd, London
Printed and Bound in Great Britain by
R. J. Acford, Chichester, Sussex

British Library Cataloguing in Publication Data

Latchem, W.E.
  GCSE chemistry for you.
  1. Chemistry
  1. Title
  540   QD33

ISBN 0-09-164591-3

Contents

Chemistry? What's that?

Chemistry is part of Science. In Science we find out about ourselves and all the things around us.

To find out we must ask questions.

Look at the picture. Pick out ten things you can see in it. Write out a list of them on paper. Can you answer questions about them?

Try these questions:

Is sea water really blue?

How could I find out if the water is sea water?

Where do clouds come from?

Steel sinks in water. Why does a steel ship float?

What moves the ship through the water?

Where does its fuel come from?

What are the cliffs made of?

What is it that makes grass green?

Why is sea water salty?

Does anyone make salt from sea water?

Why does the steel in the ship rust?

Science tries to answer these questions. It answers thousands of others too. In Science we find out about things as they are. We also find ways of changing them.

Human beings like asking questions. They began thousands of years ago. So today we know a great deal. We know about our world and ourselves. Science? There is 'a lot of it about'.

This is why Science is divided into parts.

These parts are called **branches** of Science. Chemistry is one branch. Biology is another. There are many others: Physics, Engineering, Astronomy, Electronics . . .

Each deals with a part of Science. We shall see later how they overlap.

About 4000 years ago people made metals. They built monuments like Stonehenge.

About 2500 years ago they made glass. They made cheap metals look like expensive ones. This made them think a metal like lead could be turned into gold. They tried to find something which would make it happen. This made them ask what the world is made of.

About 2300 years ago Democritus said it was made of atoms.

2000 years ago the Romans had invented central heating. They used coal as a fuel. They built straight roads. They used bronze and silver coins. They invented sewage systems.

1700 years ago the first jet engine was made!

The jet was a jet of steam.

The first Chemistry book in this country was written in 1144 AD. It was written in Latin.

In 1250 AD Roger Bacon may have discovered gunpowder. He was the first modern scientist.

Since then millions of people have worked in Science. They asked questions. They found answers. They made new materials. They invented new ways of treating disease. Men have walked on the moon. Space craft have reached the farthest planets. They have sent back photographs.

Photography is only just over a hundred years old. People travelled by balloon before aircraft were thought of. Radio and television are not many years old. Plastics are recent, too.

Most of these inventions need 'stuff'. Silver, bronze, gunpowder, medicines, coal, plastic . . . .

This is what we deal with in Chemistry.

Stonehenge is about 4000 years old.

The space shuttle Columbia, 1982.

Chemistry and substances

Our world is made of 'stuff'. It has earth, air, rocks, sand, water. . . . We are made of bones, teeth, hair, skin. . . . All this 'stuff' is called matter.

One particular kind of matter is called a substance. Iron is a substance. So is water. Salt is a substance, too. We have matter all around us. Most of it is a mixture of substances.

Matter can be solid, liquid or gas. Sand is a solid. water is a liquid and air is a gas. Solid, liquid and gas are called the **states of matter.**

Where do we get substances?

They come from the earth. For example, coal is mined. Oil is found underground. Rocks in the earth are mixtures of substances. These substances are all minerals.

The sea is a liquid, water. It has many substances dissolved in it. Salt is only one of these. Air can provide substances which are gases.

Plants give us food. They provide fibres such as cotton and linen. We use these to make clothing.

Plants provide sugar and starch. They give plant oils. They are used in making soap and cosmetics. Some of the first medicines and drugs came from plants. Plants are the source of substances like rubber.

Limestone quarrying at Sill's Quarry, Mansfield.

Drilling for oil in the North Sea.

Opencast mining for coal.

In Chemistry we find out about substances. We do this by asking three kinds of question.

1 How can we get a substance on its own? This means taking others away from it. In Chemistry we first make substances pure.

2 How do we know, or recognize, a substance? I have a white substance. How do I know whether it is chalk, salt, sugar or . . .?

3 How can we use substances to make new ones?

As an example, take the substance sugar. It is made from sugar beet or sugar cane.

Sugar beet is grown in cold climates. It is harvested. Soil is washed off. The beet is sliced and treated with hot water. The sugar passes out of the beet into the water.

The liquid is separated. Lime is added to it. This gets rid of some substances. They fall to the bottom of the liquid as sludge. The sludge is taken out by filtering.

The clear liquid is heated. It loses water. Sugar forms in it. It is removed by centrifuge. After washing it is dried ready for use.

A thick treacly liquid is left. It is called molasses. Treacle, rum and alcohol are made from it. With beet pulp it makes cattle food.

Sugar cane grows in hot climates. It is treated in the same way. The sugar formed is brown. Charcoal is used to turn it into white sugar.

So 1 We have made pure sugar.
    2 We can recognize it by its taste.
    3 We can use it to make other substances.

Using taste is dangerous. A white substance may be sugar. But it may be a poison. We need to find other ways to recognize substances.

Our world provides mixtures. From mixtures we make single substances. We make substances as pure as possible. From them we find ways of making useful new ones.

A change which gives a new substance is called a chemical change. If no new substance is formed the change is a physical one. We make the changes. We also explain them. Many questions begin with 'Why'

Harvesting sugar cane.

Copy this into your book. Fill in the blanks.

In Science we _find_ out. To do so we ask _questions_. Chemistry is a _branch_ of Science. In Chemistry we deal with _substances_. We ask three kinds of questions:

1 How can we make substances _pure_?
2 How can we _recognize_ a particular substance?
3 How can we use substances to make _new_ ones?

The world is made of 'stuff'. All this stuff is called _matter_. It can be solid, _liquid_ or _gas_. These are the different _states_ of matter.

We get substances from the _earth_. They also come from the _air_, the _sea_ and _plants_

They are mixtures.
From them we make _pure_ substances. We use these to make _new_ ones.
A change which gives a new substance is a _chemical_ change. One which gives no new substance is a _physical_ change.

find, questions, branch, substances, pure, recognize, new, matter, liquid, gas, states, earth, air, sea, plants, pure, new, chemical, physical.

Experiments

In Chemistry we find out about substances. Sometimes we can do this just by looking. Using eyes (and ears) is called observation. It may be all we need to answer some questions. To answer others we need to do tests. A test in which we do something is an **experiment**.

In Chemistry we use observation and experiment.

An experiment must be planned with care. We first decide what we want to know. What question do we need to ask and answer?

Next we work out what we need to do. Then we plan exactly how we mean to do it. The plan must say what materials we need. Last of all, we make sure the method is safe.

During the experiment watch what happens. Make notes so that you do not forget anything.

Use the results to answer the question asked.

We often do this in everyday life. Look at the drawing on the right. It shows a problem which asks a question. This is how we might work out how to answer it.

I have to deliver a newspaper at this house. The notice shows that this could be tricky. Does the dog bite?

I must give the dog a chance to bite (but not me!). I could go in wearing cricket pads. Or I could wait and watch the postman go in.

The dog might bite me above the cricket pads. I shall wait for the postman. This is safer.

The postman went through the gate. The dog rushed at him. It licked his hand.

It seems that the dog does not bite.

Does the dog bite?

An experiment may fail for many reasons.

It may have been badly planned.

I wait for the postman to come. He fails to turn up. He may already have gone. We did not know enough to plan properly. We should have asked an earlier question. What time does the postman do his round?

We may have asked the wrong question.

Perhaps the dog knows the postman well. They may be friends. Try again with a new question. **Does the dog bite strangers?**

> X. Perry, a chemist from Gwent
> Was trying to make a new scent.
> When he smelled the first pong
> He said, 'I've gone wrong!
> This isn't what X. Perry meant.'

Experiments are important. Science cannot exist without them. Planning an experiment means asking

1. What do we need to know or do?
    What is the question to be answered?

2. What shall we do to get an answer?
    Is the method we hope to use safe?

3. What substances shall we need?
    What else shall we have to use?

During and after the experiment we must

4. Watch carefully.

5. Write down exactly what happens.

6. Use these results to answer the question.

Try this!

One question on page 6 was: Is sea water really blue or is it colourless like tap water? Plan an experiment to find out.

These questions are to help you plan.
How will you get sea water?
What will you put the sea water in? A bottle? A plastic cup? A white plastic cup? A glass tumbler?
How will you take sand or seaweed from the water?

I am Willie, this is Grace.
We think experiments are ace.
We hope that you enjoy them too –
That's why it's Chemistry for You.

Plan experiments to answer these questions:

1. What is the length of a football pitch?
2. Does ice become colder when salt is added?
3. How cold must it be to freeze diesel fuel?

Hints
1. Tape measure? Pedometer? Bicycle wheel?
2. How do we measure hotness and coldness?
3. How shall we cool the fuel? What shall we put it in? Is the fuel safe to use?

Try this – how to find out

Sea water comes only from the sea. It can be carried best in a clean bottle. Strain it through cloth to remove seaweed or sand. A white cup makes it easy to see colour.

Safety

In Science we find out about ourselves and our world. To do this we ask questions. Those in Chemistry are about substances. Sometimes we can answer them just by looking. Using eyes (and ears) is called observation.

To answer other questions we need to do tests. A test in which we do something is called an experiment. It must be carefully planned. Even then it may be dangerous. So may the substances we use in doing it. We need to make Science safe for ourselves and others.

Chemistry is like the dog. There is a chance that it may be dangerous. Always follow the simple rules given below.

Safety rules

Experiments marked * should only be done by a teacher.
Keep working areas clear of cases, books etc.
Listen carefully to instructions.
Work with care and watch results.
Remember that clothes and hair will burn.
Point test tubes away from yourself and others.
Wear eyeshields if there is any risk to eyes. (To be quite safe wear them all the time.)
Report any accident, however small, at once.

DO NOT run, push or be silly in a laboratory.
DO NOT taste anything without permission.
DO NOT smell a gas except with great care.
DO NOT try your own experiments without permission.

Willie, Grace and George appear in this book:

Willie, for everyone's protection,
Points test tubes in a safe direction.
Grace, when stuff from test tubes flies,
Has eyeshields to protect her eyes.
George treats the bunsen flame with care,
He wouldn't like to lose his hair.
In all experiments we do in schools,
We must obey the Safety rules.

The drawings shown are labels. They are used for dangerous substances. Each picture explains what the danger is. They appear on bottles and containers. Look for them on tankers and trucks.

explosive

toxic (poisonous)

corrosive

oxidizing agent (oxidant)

flammable (will burn)

harmful or irritating

The disaster at Bhopal. Over 2000 people died after a huge leak of poisonous gas.

Summary

The world is made of 'stuff'. This stuff is called **matter.** Some matter is solid. Some is liquid and some is gas. Solid, liquid and gas are the three **states of matter.**

A substance is one particular kind of matter. Iron, water, sugar, salt are all substances.

Substances are made from the matter around us. They come from the earth, the sea, the air and from plants. They are mixtures of substances.

From them chemists make pure single substances. They must know how to recognize a substance. We can recognize substances by taste. But this is very dangerous. Other methods are needed.

Chemists make new substances from those we have. A change which gives a new substance is a **chemical change.** If no new substance is formed the change is a physical one.

Chemists find out by observation. They also answer questions by experiments. These must be planned with care. Above all, the method must be a safe one. SAFETY IS VERY IMPORTANT.

Things to do

Look at the labels on substances at home. Make a list of those with 'Pure' on the label. Look for the warning signs shown on this page.

Questions

1. Explain what we do in Science. Why is Science divided into branches such as Chemistry?
   What parts of Science does Chemistry deal with?

2. Your father has been reading your Chemistry book. He wants to know what these words mean. Tell him. Use one sentence for each word.
   a) matter, b) substance, c) chemical change, d) physical change, e) observation, f) experiment.

3. What branches of Science deal with
   a) living things,
   b) stars, planets and other heavenly bodies,
   c) changes which form no new subsances,
   d) the treatment of disease?

Making substances pure

We get substances from the air, the sea and plants. They also come from the earth. One of these substances is rock salt.

Drop a lump of rock salt on the bench. You can hear how it gets its name. Observation means using ears as well as eyes.

A machine cutting rock salt in a Cheshire mine.

Experiment 1.1 What does rock salt contain?

Put a small lump of rock salt in a mortar. Crush it to a powder with a pestle. Look at it through a magnifying glass. Try a microscope as well.

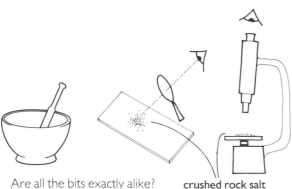

crushed rock salt

Are all the bits exactly alike?
Are they all the same colour?
Are they all the same shape?
Do any of the bits look like sand?

Observation shows that rock salt is a mixture. Most of it is salt. Some bits look like sand.

To get salt from rock salt we must take away the sand. How? Make use of some way in which they differ.

Two nuts fall off your bike. It happens on the beach. Get them back. How? Make use of some way in which nuts differ from sand.
Pick them out by hand? Nuts and sand look different.
Put the sand through a sieve? Nuts are bigger than sand grains.
Run a magnet over the sand? Steel moves towards a magnet.

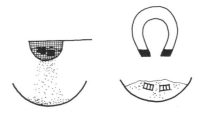

What difference is there between grains of sand and salt?
Can we pick sand grains out with tweezers?
Salt dissolves in water. Does sand?

14

Experiment 1.2 To separate the sand from the salt

Take a small beaker. Put in half a spoonful of crushed rock salt. Add water to a depth of 1 cm.

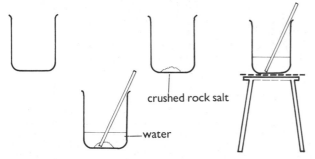

crushed rock salt

water

Heat the beaker on a wire gauze on a tripod. Stir the water with a glass rod. Stop heating when the salt has gone.

Fold a filter paper as shown. Put the cone into a funnel. Put an evaporating dish under it. Pour the liquid from the beaker into the cone.

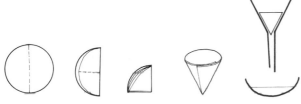

Where is the sand and where is the salt?

The paper cone **filters** the liquid. The liquid which runs through it is called the **filtrate**. The salt is in the filtrate. The salt **dissolved** in the water. A substance dissolved in a liquid is a **solution**. In the dish is a solution of salt in water. The sand is trapped in the filter paper cone. Pour a little water over it. This will wash any salty water out of it.

How can we get salt from this solution?
What will happen if we heat it in the dish?

Pure salt crystals.

Remember to wear eyeshields.

Experiment 1.3 To get salt from the solution

Put the dish on wire gauze. Heat it gently on a tripod. Stop when most of the water has gone.

Take up some of the hot liquid in a dropper. Put drops on a glass slide. White bits should form in the drops. If not, heat the dish again. Look at the bits with a microscope. Taste them. (It is safe!)

Do the white bits in the liquid taste of salt?
Is there any sand mixed with them?
Are they like those shown in the photograph?

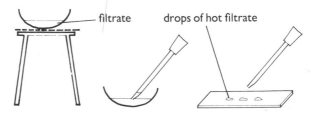

filtrate

drops of hot filtrate

Heating the solution makes it lose water. It comes out of the dish as steam, or water vapour. This is called **evaporation**. Some water is left. It may not be enough to dissolve all the salt. Some bits of salt come out of solution. They have a definite shape, like cubes. They are called cubic **crystals** of salt.

Solutions

Salt put into water disappears. It spreads through the water. Each drop of water has salt in it. Salt has dissolved in water.

Any substance dissolved in a liquid makes a solution. The liquid is called the **solvent**. The dissolved substance is the **solute**.

Experiment 1.4 To find out more about solutions

1. Half fill a test tube with distilled water. Put in a little salt. Stopper the tube. Shake the mixture for a minute or two.

Add more salt. Stopper the tube and shake it. When no more salt dissolves, add a little more. Heat the tube gently. Find out if more salt dissolves in the hot water.

Put drops of hot solution on a glass slide. Watch them through a microscope as they cool.

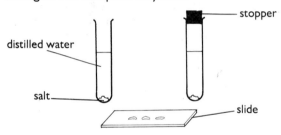

Now try copper sulphate instead of salt.

    Do both substances dissolve in water?
    Is there a limit to how much will dissolve?
    Why do crystals form when hot solutions cool?
    Salt and copper sulphate crystals differ. How?
    Does hot water dissolve more solute or less?

2. Put distilled water in a test tube. Add a little chalk. Heat the tube gently. Filter the mixture. Put drops of filtrate on a slide. Heat them as shown in the drawing.

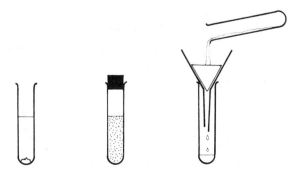

George dreamed of a job in the city.
He put rock salt on chips. What a pity!
He first should have planned
To take out the sand.
Then the chips taste just salty, not gritty.

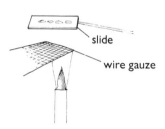

Does chalk appear to dissolve in water?
The filtrate runs through the paper. Is it clear?
Is any solid left on the glass slide?
Does chalk dissolve in distilled water?

1. If enough salt is added to water some will not dissolve. There is a limit to how much solute will dissolve. If no more solute will dissolve the solution is **saturated**. More solute dissolves in hot water than in cold. Cool a hot saturated solution. It will hold less solute. Some solute will be thrown out. It may appear as crystals.

2. Heating the filtrate leaves no solid. Chalk does not dissolve.

Other solvents and solubility

Experiment 1.5 To try other liquids as solvents

Put out all bunsen burners. Put ethanol (meths) in four test tubes. In one put a small amount of salt. In the second put a little sugar. Add some copper sulphate to the third. Stopper the tubes. Shake them. Heat them by standing them in hot water.

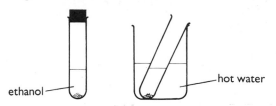

ethanol — hot water

Shake an iodine crystal in the fourth tube. Add two more crystals. Drop three similar crystals into the same volume of water. Shake both tubes.

> Which solids dissolve in ethanol?
> Which dissolve in both ethanol and water?
> Does iodine dissolve better in ethanol or in water?

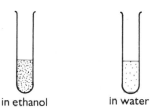

in ethanol   in water

Salt and copper sulphate dissolve in water. They do not dissolve in ethanol. Sugar dissolves in both. So does iodine. Its solutions are brown. Much less dissolves in water. This solution is much paler.

Any liquid can act as a solvent. Some substances dissolve. They are **soluble** in the solvent. Salt is **soluble** in water. Iodine is **soluble** in ethanol.

Others do not dissolve. They are **insoluble**. Salt is **insoluble** in ethanol. Chalk is **insoluble** in water.

Solute and solvent make a solution. If the amount of solute is small the solution is **dilute**. If it is large the solution is **concentrated**. If no more will dissolve the solution is **saturated**. Hot solvents usually dissolve more solute than cold ones.

These experiments show that solids dissolve. Can gases and liquids dissolve? We shall use ethanol and oil as liquids. The simplest gas to use is air.

Experiment 1.6 To find out if liquids and gases dissolve

1. Half fill two test tubes with water. Add oil to one. Add ethanol to the other. Stopper the tubes. Shake them vigorously. Allow them to stand.

*2. Take a round-bottomed flask. Fill it nearly full with water. Set it up as shown. Connect a filter pump to it. Draw air through the water for fifteen minutes.

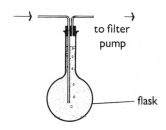

to filter pump

flask

Fill the flask to the top. Fit it with a stopper and tube. Set it up as shown. Fill a gas jar with water. Turn it upside down in the trough. Heat the flask. Let any gas in the tube escape. Hold the gas jar over the tube.

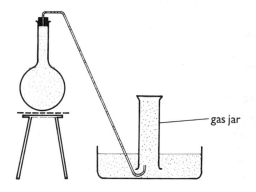

gas jar

> Does oil dissolve in water?
> Do they form separate layers in the tube?
> Does ethanol dissolve in water?
> Does water dissolve in ethanol?
> Does gas come out of the heated water?

Oil and water form separate layers. Oil does not dissolve in water. Water does not dissolve in oil.
Ethanol and water form one liquid. Each liquid will dissolve in the other.

Bubbles rise in the heated water. They pass through the delivery tube. Gas collects in the gas jar. It is air. It must have been dissolved in the water.

Liquids and gases can dissolve in solvents.

Salt of the earth

When rain falls it runs through soil and rocks. Substances dissolve in the water. This water runs into streams and rivers. More substances dissolve. The rivers carry them to the sea.

This means that sea water has many substances dissolved in it. The main one is salt. An area of sea is sometimes cut off. It becomes an inland sea. One example today is the Dead Sea. Rivers flow in; nothing flows out. It evaporates like the salt solution in our dish.

Millions of years ago there were inland seas. One covered the middle of the British Isles. The climate was hot. The sea lost water. When all the water had gone, salt was left. It was not pure. It had dust and other substances in it.

The salt became covered with dust. Other changes turned the dust into rock. So the salt layers are now found underground. The photograph on page 12 shows rock salt being mined in Cheshire. Substances in the earth are called **minerals**.

Not much salt is mined. Most of it is brought to the surface in a different way. Water is pumped down to the salt layer. The salt dissolves. The solution, called brine, is pumped up. It is evaporated to give salt crystals.

Salt is also made from the sea. Sea water is run into shallow 'pans'. In hot climates it loses water quickly. Sea salt is formed. It contains other substances besides common salt.

About 80 million tonnes of salt is used each year. Animals cannot live without it. Human beings need it too. We use it to flavour food.
Salt preserves food, too. Food treated with salt is 'cured'. Cured meat and fish keep longer.

Salt is used in making other substances such as soap, weed killers, household bleach and soda.

Chemists make salt and other substances pure. They find out about them and use them to make new substances.

Cheddar Gorge, cut out of the rock by water.

A salt pan in Ethiopia.

Pure substances

A substance used in Chemistry should be pure. If not, we get wrong ideas about it. Suppose the only salt we knew was rock salt. We would think that salt is brown, salty and gritty.

Most substances used in Chemistry are kept in glass or plastic bottles. Each has a label. It gives the chemical name of the substance. The chemical name of common salt is sodium chloride.

This is the label of a salt bottle. It is much purer salt than we made. The label tells us how pure it is. 100 grams of this substance contain at least 99.5 grams of salt. It still has other substances in it. The label names them. The amounts are very small indeed.

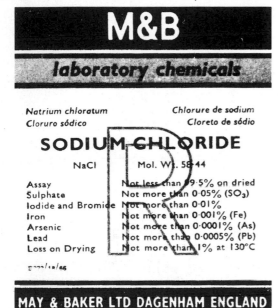

M&B
laboratory chemicals

Natrium chloratum   Chlorure de sodium
Cloruro sódico   Cloreto de sódio

SODIUM CHLORIDE

NaCl   Mol. Wt. 58·44

Assay	Not less than 99·5% on dried
Sulphate	Not more than 0·05% (SO₃)
Iodide and Bromide	Not more than 0·01%
Iron	Not more than 0·001% (Fe)
Arsenic	Not more than 0·0001% (As)
Lead	Not more than 0·0005% (Pb)
Loss on Drying	Not more than 1% at 130°C

MAY & BAKER LTD DAGENHAM ENGLAND

The label from a bottle of the purest salt.

In Science then, 'pure' has one meaning. A pure substance is a single substance. We can never make it 100% pure. We can only get close to it.

In daily life, pure has a different meaning. We buy 'pure honey'. We eat 'pure butter'. A clothes tag may say 'pure wool'. Wool, honey and butter are mixtures. Not one is a single substance. Pure means nothing has been added or removed. The honey is as bees made it! The sock has only wool in it!

Foods have many substances added. Man-made dyes, sweeteners and flavours are examples. They are 'food additives'. Read the label on the food!

Summary

Substances found in the earth are called minerals. Rock salt is a mineral. It is a mixture. It has sand and salt in it. Salt dissolves in water. We say it is soluble. Sand is insoluble. It does not dissolve in water. We filter the solution to remove the sand.

Heating the salt solution makes it lose water. This is evaporation. The water gets less. In the end the solution is saturated. Some of the salt comes out on cooling. It forms cubic crystals. Other substances are left in the water.

All liquids are solvents. Solutes dissolve in them. A fixed amount of solvent can dissolve only so much solute. A saturated solution can dissolve no more solute. Most solids dissolve more in hot solvent than in cold. Gases and liquids dissolve in solvents. Gases escape on heating the solution.

Questions

1. In one sentence for each, explain the meaning of these words: solvent, solute, solution, soluble, saturated, crystals.

2. Work out a method of getting back
   a) iron nails lost in long grass,
   b) sugar which has fallen into chalk,
   c) iodine mixed with salt.

3. There is a muddy, weedy pond in your garden. How would you
   a) get clear water, free from mud and weed,
   b) find out if air is dissolved in the water,
   c) find out if any solid is dissolved in it?

4. You have four beakers. Each contains a liquid. They are
   a) sea water,   c) water with dissolved air,
   b) rain water,   d) copper sulphate solution.
   Invent a way of showing which is which. How could you find out if solution d) is saturated?

Things to do

Make a list of foods which are cured with salt. Find out the crystal shape of salt in your salt cellar. Use a magnifying glass.

Find out if tap water is a solution. Leave some on a saucer on the radiator, or heat some on a glass slide.

Look at food labels. List those which use the word 'pure'. Which contain additives? Some carry a date. Why?

Other substances but mainly water

Black ink is a common liquid. It may be a single black substance. It may be a solution of a black solid in a solvent. How can we find out?

Heat it. A black liquid evaporates. Nothing is left. A black solution evaporates, too. However, only the solvent boils away. The black solute will remain at the end.

Experiment 1.7 To find out about ink

Gently heat black ink in an evaporating dish.

The ink boils. It seems to give off steam. A black solid is left in the dish. Ink is a solution. It seems to be a black solid in water. How shall we know if the solvent is water?

What do we know about water? It has no colour or taste. When heated it turns into steam. When cooled it turns to ice.

Experiment 1.8 To find out more about water

Work in pairs. Fill a large test tube half full of ice. Hold it in a clamp and stand. Support a thermometer in the ice. Take its temperature.

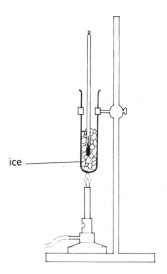

ice

Gently heat the tube. Use a flame about 2 cm high. One partner looks at a watch. Every ten seconds he (or she) says 'Now'. The other calls out the temperature. The partner with the watch writes it down. Change over when the temperature is 50 °C.

Heat the water until it boils. Then take the temperature. Heat it more strongly. Find out if the temperature changes. Put in some salt. Take the boiling temperature again.
Add salt to fresh ice. Take the temperature.

Write down your results:
Melting ice was at _____ °C.
Ice and salt were at _____ °C.
Salt makes ice melt _____ quickly.
Boiling water was at _____ °C.
When strongly heated it boils at _____ °C.
Salt solution boils at _____ °C.

Ice is at 0 °C until most of it has melted. 0 °C is its melting point.
Ice and salt fall below 0°C. So the ice melts faster.
Water boils at 100 °C. 100 °C is its boiling point. Stronger heating does not alter it.
Adding salt makes the boiling point go up.

Properties

Is this man an aircraft pilot? He looks like one. But can he do what a pilot does? Can he fly a plane? Anyone can look like a pilot. All he needs is a uniform. The real test is to ask him to fly a plane.

We know people by what they look like: their appearance.
We also recognize them by what they do: their behaviour.
What they can do is the better test.
Behaviour is more important than **appearance**.

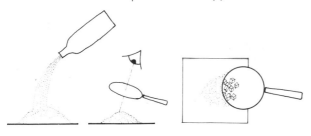

Is this substance salt? It is white. Its crystals are cubic.
Its appearance is right.
What about its behaviour?
It tastes salty. However, tasting is dangerous. We need to know more properties of salt.

The properties of a substance are its appearance and behaviour.

There is a liquid in the beaker. Is it water?
What are the properties of water?
Pure water has no colour, no smell, no taste.
It melts at 0 °C. It boils at 100 °C.

Which are the best properties to use? The most useful properties are those which depend on what water can do. Of course, the water must be pure. Dissolved substances alter its boiling point. A change of pressure will alter it, too.

We have made pure salt from a salt solution. Can we get pure water from a salt solution?

When salt solution boils it turns into steam. The salt is left behind. How can we turn the steam back to water?
Three ways are shown in the drawings. Will any of them work? Which is the best method?

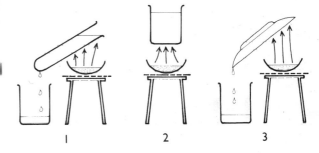

I	2	3

Look at the picture. Tell me how
To prove this animal's a cow.
Its horns? Its udder? I would choose
To show that it gives milk and moos.

Try to tell me how I oughter
Find out if a liquid's water.
Its colour? Taste? No, let it freeze
And show it melts at 0°.
(Celsius, of course)

Pure water

To get water from a salt solution it must be boiled. The water turns into steam. Cooling the steam turns it into water again. On the last page were three methods of cooling. The best is method 3. It is the best way to collect the water.

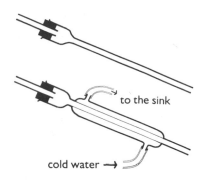

However, the plate soon becomes hot. It no longer cools the steam. A German named Leibig found a better method. He passed the steam through a glass tube. This has a second tube round it. Cold water passes through this outside tube. It cools the steam **all the time**.

to the sink

This is called Leibig's condenser.

cold water →

*Experiment 1.9 To make pure water

This is a distilling flask. Put salt solution into it.

Add a drop of ink. Put a thermometer into the flask.

Fit a condenser to the side neck of the flask.
Fit rubber tubing to top and bottom of the condenser.
Connect the bottom to the tap. Turn on the tap.

Cold water fills the outside tube. It runs away to the sink.

Boil the solution in the flask.

Steam passes into the condenser. Cold water in the outside tube cools it. It turns back to water which is collected.

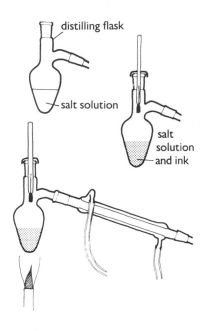

distilling flask

salt solution

salt solution and ink

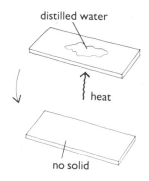

distilled water

heat

no solid

Is the water coloured?
Does it taste of salt?
What does the thermometer show?
This is the temperature of steam. Is the water pure?

This method is called distillation. The water formed is called distilled water. It boils at 100 °C and is pure. It has no colour. It does not taste of salt. The salt and the black dye are left behind in the distilling flask.

Note that it has no **solids** in it. If we heat it on a glass slide it will boil away. No solid will be left on the slide.

All liquids are solvents. They dissolve solids. They can also dissolve other liquids. Ethanol will dissolve in water, for example.

Suppose we boil this solution. **Both** liquids will turn into vapour. Some of each will be given off. A condenser cools the vapour, giving a liquid mixture.

Copy this into your book. Fill in the blanks.

We know people by what they look like. This is called their _____. We also recognize them by what they can do. This is called their _____.

We recognize a substance in the same way. The appearance and behaviour of a substance are called its _____. Its _____ is a more useful property than its _____.

To show its true properties a substance must be _____. Water boils at _____ °C. Salt dissolved in it makes the boiling point _____. Ice melts at _____ °C. This temperature is its _____ point. Adding salt makes it _____.

Melting point and boiling point are useful _____. They are used to find out if a substance is _____.

When a liquid is heated it gives off _____. Cooling this turns it back into _____. When water boils it gives _____. The best cooling method is Leibig's _____. It has two tubes. Steam passes through the _____ one. Cold water runs through the _____ one. It is run into the condenser at the _____.

Heating a liquid and cooling the vapour is called _____. The condensed steam is called _____ water. It is _____. If pure water is boiled away, no _____ is left.

The engine roared. It wasn't nice!
The wheels kept slipping on the ice.
Willie knew what was at fault,
Under the wheels he scattered salt.
The car moved off! It didn't falter.
Salt had melted ice to water.

Questions

1. A wet road soon becomes dry. Does it dry more quickly in hot or cold weather? Where does the water go? What does it turn into? Water vapour rising into the air cools. What will it form? What are clouds made of? Where does rain come from?

2. Sea water evaporates. Will the vapour contain salt? Will rain drops contain dissolved salt? How pure is rain water? Can it be used instead of distilled water?

3. It is a cold night. The air temperature is −1 °C. There is ice on the road. Driving on ice is dangerous.

What is the melting point of ice?
Is the air hotter than this?
Will the ice melt in the air?
Where does the ice come from?

Salt is put on the ice. Look at the thermometer.
What is the melting point of ice and salt?
Is the air hotter than this?
Will the ice melt now?

4. You are stranded on a desert island. You have fuel and kitchen vessels. You have no salt or drinking water. Invent a method of getting both.

5. A beaker contains 100 grams of water. It is at 20°C. 36 grams of salt are dissolved in it. This is a saturated solution. What does saturated mean?
   The solution is heated until it boils.
   a) Will more salt dissolve in it?
   b) Is its boiling point 100°C?
Half the water is boiled off. The rest cools to 20°C.
   c) How much salt comes out of the water?
   d) How can we separate the salt and water?

Separating other mixtures

Ink

Ink is a mixture. Black ink is a black solid in a solvent. Is the black solid a single substance? Or is it a mixture?

Experiment 1.10 To find out if the black solid is a mixture

Take a 10 cm square of white blotting paper. Stand it on a beaker or a petri dish. Put one drop of black ink in the middle of the paper.

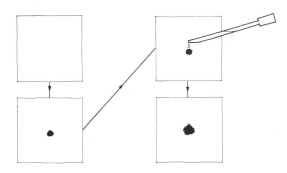

Fill a dropper with water. Put one drop of it on the ink blot. Let it soak in. Put a second drop in the middle of the blot. Let this soak in. Add drops until the colour gets to the edge of the paper.

What happens?
Is the result like the drawing?
How many colours could you see?
Is the black solid a mixture?

Try this method with inks of other colours.

The water soaks in. It spreads out through the paper. It takes the colour with it. Some coloured dyes are carried further than others. So each will end up as a ring of colour.

Black ink gives blue, orange and yellow. Did you see any other colours?

The method is called chromatography (crow-mat-og-raffy). The ring pattern is called a chromatogram (crow-mat-o-gram). It comes from Greek words meaning 'colour writing'.

The method was first used for coloured dyes. It can be used to separate substances with no colour. Can you see any problem with these?

Plants

Plants contain many coloured substances. Grass is green. Flowers and fruit are coloured. Green leaves go brown. How do we know that the coloured substances do not dissolve in water?

In wet weather grass is still green. The green dye does not dissolve in cold water. We need to find another solvent.

Experiment 1.11 To test the green substance in grass

Cut grass into short pieces. Put it in a mortar. Add some ethanol (meths). Grind the grass in it with a pestle. Pour the liquid into a small test tube. Let it stand or use a centrifuge.

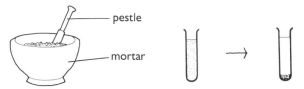

Collect the clear liquid in a dropper. Make two cuts in a filter paper as shown. Put one drop of the green solution in the middle of the paper. Let it dry. Add a second drop. Let this dry. Add a third drop on the same spot.

paper tongue

Put ethanol into a petri dish. Put the paper on the dish. Bend the paper tongue so that it dips into the ethanol. Leave it.

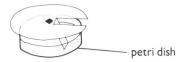

petri dish

How does the ethanol reach the green dye?
Does this method work as well as using a dropper?
How many coloured rings did you see?
Is the green substance a mixture?

Ethanol dissolves the green substance in grass. The solution has bits of grass in it. These slowly settle to the bottom. We used three drops of solution. This gives more green substance than one drop. The colours are less faint.

There are three rings of colour. The outer one is orange. This orange colour is xanthophyll. The middle ring is green chlorophyll. You may see a red inner ring of carotene.

Using chromatography

Chromatography separates coloured substances. It can be used to find out what a mixture contains. The drawings show a third way of making a chromatogram.

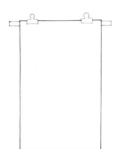

First get a piece of blotting paper (or filter paper) and hang it on a glass rod with two clips.

Use a solution of the mixture. Put one drop on the paper (U). Choose substances which may be in the unknown mixture. Put drops of their solutions on the bottom of the paper (A, B, C, D, E).

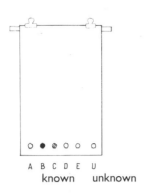

A B C D E U
known   unknown

Now hang the paper in a container with some solvent at the bottom. The paper should dip into the solvent.

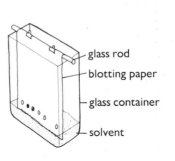

- glass rod
- blotting paper
- glass container
- solvent

side view

The solvent rises up the paper. It carries the known dyes with it. The unknown is carried, too. Look at the results. The unknown gives two dyes. They look like B and C. They have been carried to the same height, too. Unknown dye = B + C.

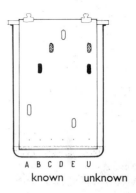

A B C D E U
known   unknown

Substances with no colour behave in the same way. The problem is that we cannot see them. We can find where they are. We simply turn them into coloured substances. A liquid spray does this.

Powders may be used instead of paper. Chalk is an example. It is put into a long tube. The mixture is added at the top. Solvent is put in. It carries the dyes down the tube. The mixture is separated. We can find out what it contains.

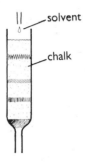

- solvent
- chalk

The centrifuge

This is a simple centrifuge. It is for use with test tubes. Notice the test tube holders.

The idea of a centrifuge is simple. Anything swung in a circle is thrown outwards. The same idea is used in a spin dryer. Wet clothes are put in a container with holes in it. It is rotated at high speed. Clothes and water are thrown outwards. The water goes through the holes. The clothes do not.

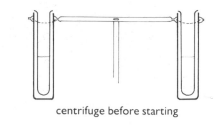

centrifuge before starting

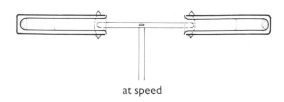

at speed

Centrifuges are used in industry. In design they are more like spin dryers. They can separate cream from milk. They separate salt crystals from a solution. They do what filtering does, but they do it much faster.

Summary

The colour in ink is a dye. A drop of it is put on paper. A solvent is added. This washes the dye outwards. The paper holds some dyes more tightly than others. They will not travel so far. The method separates coloured dyes.

This method is called chromatography. The paper can be held upright as well as flat. Powders such as chalk can be used instead of paper. Other solvents are used as well as water. The result is called a chromatogram.

The method can be used for any substances. But colourless ones cannot be seen. Some way of finding out where they are is needed. They are turned into coloured ones.

A centrifuge takes solids out of liquids. The mixture is whirled round at high speed. The solid is thrown outwards. This takes it to the bottom of the liquid. The liquid is poured off. In other centrifuges liquid passes through the sides of the container. Solid is held inside.

A centrifuge.

A centrifuge is used to separate solid from liquid. A laboratory one is shown in the photograph above.
On each arm it has test tube holders.
They swing on pivots. The test tubes put into them contain the mixture.

The arms rotate at high speed. Holders and test tubes are thrown outwards. So is the solid.
It ends up at the bottom of each tube.

Centrifuges separating sugar crystals.

Things to do

1. Use ethanol to dissolve coloured substances out of flower petals. Find out if they are mixtures or single substances.

2. Test the dye in Smarties.
   Warm six of them in a little water. Pour off the coloured solution. Boil white wool in 1% ammonia solution. Wash it in water. Put a metre length in the coloured solution. Add a little ethanoic acid. The wool takes up the dye.

   Remove the wool. Put it into a little 1% ammonia solution. Let the dye come out. Boil the liquid away. Dissolve the solid in one drop of water. Use this in a chromatography experiment. The solvent is: 3 parts butan-1-ol, 1 part ethanol, 1 part of 1% ammonia solution.

Questions

1. Look at the drawing. The two tubes are full of powder. In one, a known mixture of substances has been used. They are A, B, C, D and E. An unknown substance was used in the other. The same solvent was used in each.

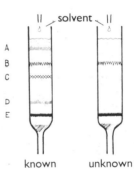

   a) Name one powder which could be used in the tubes.
   b) Which substance moved furthest through it?
   c) Which coloured band has moved least?
   d) How many bands are there in the second tube?
   e) Is the unknown substance a mixture?
   f) What does it contain?

2. Explain how you would separate these mixtures:
   a) mud and water,
   b) chalk and salt,
   c) tea leaves and tea,
   d) chalk and water,
   e) brass nails and iron nails.
   f) sugar and salt,
   g) the coloured dyes in flower petals,
   h) iron nails and aluminium nails,
   j) a mixture of ethanol and water.

27

Separating liquids

Crude oil

We get many substances from the earth. One of these is crude oil. It is also called petroleum.

Petroleum is found in some kinds of rock. The rock is porous. It holds the oil just as a sponge holds water. Above and below it are rocks which are not porous. Oil is found in rocks under the sea as well as under the land.

A hole is drilled down to the oil. Pipes are fed into the hole. Gas may exist above the oil. It may force the oil up pipes. In other wells the oil must be pumped up. It contains hundreds of substances. It is taken by pipeline and tanker to a refinery. Here it is distilled.

Experiment 1.12 To distil crude oil

Use a test tube with a side neck. Fill it to a depth of 2 cm with crude oil. Push rocksill into the oil. This soaks up the liquid.

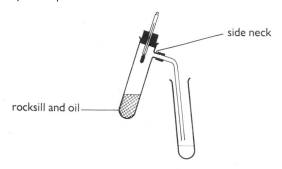

Stopper the tube. Use a thermometer in the stopper. Heat the oil gently. Stop when the thermometer shows 70 °C. Stopper the collecting tube. Replace it with a second tube.

Heat the oil again. Collect liquid up to 120 °C. Change the collecting tube at 120 °C, 170 °C and 200 °C. Put a stopper in each tube.

Drilling for oil.

Test the liquid in each of the four tubes.
1. Tilt the tube. Find out how runny each liquid is.
2. Put a few drops on a watch glass. Leave it.
3. Put drops on a watch glass again. Try to light it.
If a liquid does not burn, put a cotton wool wick in it. Try to light the wick. Each liquid is called a 'fraction'.

watch glasses

Are the fractions coloured?
Which of them is the least runny?
Which fraction evaporates most quickly?
Which burns best and which is worst?

There are many substances in crude oil. Some have low boiling points. They boil off first. They are the first fraction. Higher boiling point liquids come next.

28

Refining crude oil

A crude oil distillation unit.

The rising vapour passes through trays. These have bubble caps. They force the vapour to pass through liquid already formed.

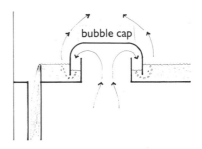

This cools the vapour. Some turns back to liquid. This liquid collects on the trays. It is run off.

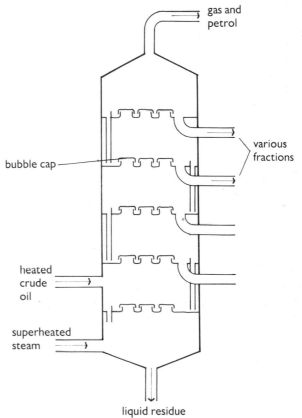

The first fraction is runny. It evaporates fast. It burns well. It is not one substance but a mixture of many.

The second fraction is much the same. It evaporates less quickly. It burns just as well. It is a mixture.

The third and fourth are less runny. They evaporate slowly. The third burns with a smoky flame. The fourth needs a wick.

This is fractional distillation. A refinery deals with crude oil in the same way. But the method has to be continuous. Hot crude oil is passed into a tall tower. Superheated steam keeps it hot. Vapour rises and cools. Liquid falls.

Vapour of liquids with low boiling points goes on. It turns back to liquid higher up the tower. This can also be run off. The crude oil is separated into fractions. The method goes on continuously.

We shall study this in more detail later in the course.

Making new substances

Chemists make pure substances. They find out
what their properties are. They find ways of making
new substances from them. One method of doing
this is to heat substances we already have.

Experiment 1.14 The effect of heating wood

Use pieces of wood as long as a match but thicker. Hold
one piece in tongs. Put it in the hot part of a bunsen
flame. Take it out when it burns.

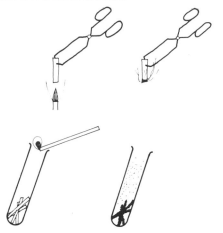

Put four pieces in a test tube. Hold it at the open end in
a holder. Heat the wood. Put a lighted splint at the
mouth of the tube. Stand the tube on a tile to cool.
Take out the black pieces. Try to write with one. Put
another in the bunsen flame.

Does the wood burn with a flame in the air?
Does it burn with a flame in the test tube?
Have new substances been formed in the tube?
Does the black solid write and burn?

Experiment 1.15 The effect of heat on other substances

  Coal
  Paraffin wax
  Chalk
  Salt
*Iodine
*Red lead
  Copper sulphate

Use a test tube for each. Look at each substance before
heating. Look at what is left after heating. Describe any
changes. Light a splint. Hold the flame at the mouth of
each tube.

Do any of the heated substances burn?
Does anything come out of the tube?
Which tubes have new substances in them?
Which are the same after heating as before?
What happens to the lighted splint?

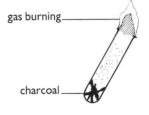

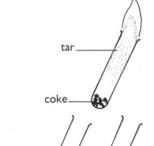

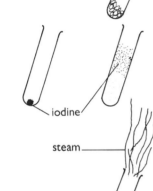

gas burning

charcoal

tar

coke

wax

iodine

steam

white powder

Wood begins to burn in the flame. It burns with a flame when taken out. A small amount of grey ash is left.

In the test tube the wood does not burn. There is no flame in the tube. The wood turns into black sticks. They make a soft black mark on paper. They also burn.

A dark brown tar forms in the test tube. Something burns at the mouth of the tube. The splint sets it on fire. We can see nothing coming out. So it must be a gas.

wood heated → charcoal + wood tar + a gas which burns

Coal gives the same kind of result as wood. The black lumps left in the tube are coke.

coal heated → coke + coal tar + a gas which burns

Paraffin wax melts to a liquid. It turns back to wax on cooling. No new substance is formed.

Chalk and **salt** No change can be seen in either substance.

Iodine A black crystal of iodine turns into a violet substance. This is iodine vapour. No liquid is seen. As it goes up the tube the vapour cools. It turns back to shiny black crystals.

Red lead The red powder changes to a yellow one. The flame of the splint burns brighter. This means something is coming out of the tube. Nothing can be seen. It must be a gas.

red lead heated → a yellow substance + a gas

Copper sulphate crystals The blue crystals slowly turn white. Steam comes out of the tube. The splint goes out. Drops of liquid form in the tube.

copper sulphate crystals → a white powder + steam (?)

Heating does not change some substances. They just get hotter. Some, like wax, get hot enough to melt. Cooling reverses the change. The liquid turns back to solid.

Heating some substances forms new ones. Wood and coal break down into many new substances. The yellow solid from red lead may be a new substance. A change of colour does not always mean a new substance. A gas given off does. The white substance from copper sulphate may be a new one.

**A change which gives new substances is a chemical change.
If no new substance is formed the change is a physical one.**

Heating metals

Write down the names of all the metals you know.

How do you know that they are metals?
What are the properties of metals?
Do they all have a shiny surface?
Do they float or sink in water?

Check your list with the one on the next page. Give yourself a mark for each metal in your list.

All metals have a shiny surface. If it is not shiny a metal can be polished. The shine is called **metallic lustre**. It is quite different from the shine on plastic or glass.

Metals can be recognized by appearance. But behaviour is more important. So we shall find out what metals can do.

Experiment 1.16 To heat metals (wear eyeshields)

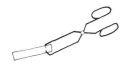

1. Clean a strip of copper with emery paper. Hold it in tongs. Heat it gently. Leave it on a tile to cool.

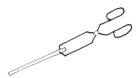

2. Do the same with a 1 cm strip of magnesium.

platinum — glass

3. Take a platinum wire sealed into glass. Use the glass as a holder. Heat the wire strongly.

crucible lid

*4. Rub a small lump of lead with emery paper. Put it

on a crucible lid. Put the lid on a pipe clay triangle, on a tripod. Heat it gently, then strongly.

iron wire

5. Rub iron wire with emery paper. Hold it in tongs. Heat it.

Which metal is exactly the same after heating?
In what ways have the other metals changed?
Which metal burns with a bright, white flame?

All the metals get hot. Platinum becomes white hot. It does not melt. When it cools it is the same as before.

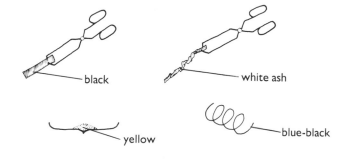
black — white ash
yellow — blue-black

Magnesium burns with a bright, white flame. The result is a dull white powder. It is not a bit like magnesium. Copper becomes coated with a black powder. Lead melts. It slowly changes to a yellow powder. Iron is a grey metal. It forms a blue-black substance.

Summary

Heating makes substances get hotter. Some, like wax, become hot enough to melt. Cooling will reverse the change. No new substance is formed. This kind of change is called a **physical change**. Such changes are easy to reverse.

Heating melts ice to water. Cooling turns the water back to ice. No new substance is formed. Ice to water, water to ice are physical changes.

Heating changes some substances into new ones. Heating coal splits it up into new substances. Shiny magnesium metal burns. It leaves a white powder. This is very different from the metal. It seems likely to be a new substance.

This kind of change is called a **chemical change**. It is not easy to reverse. A chemical change forms at least one new substance.

The new substance may be a gas. Most gases cannot be seen. We can collect them. Most of them affect a lighted splint. Some have a smell.

Metals have a shiny surface. It is called metallic lustre. Heated metals often lose it. They become coated with a coloured solid. This is likely to be a new substance. Colour changes often show a new substance formed.

Mad to light magnesium wire,
Willie sets the house on fire.
Willie's wallet, full of cash, is
Lost in the fire and burnt to ashes.
Though all his notes have gone, it's strange –
Willie has managed to get some change.

Questions

1. Look at each common change listed below. Decide if a new substance is formed. Decide if the change is easy to reverse. Say whether the change is a chemical or a physical one.
   a) milk going sour,
   b) iron rusting,
   c) boiling water turning into steam,
   d) frost forming on a cold night,
   e) a wet road drying in the wind,
   f) a forest fire,
   g) salt dissolving in water,
   h) clouds forming in a clear sky,
   j) bread turning into burnt toast.

2. Copy this into your book. Choose the best answer from the ones in brackets. Put this answer in.

   New substances are formed by (physical/chemical/simple) changes. Heating wood in air makes it (hot/burn/boil). It turns into a (brown/grey/yellow) ash. This (is/is not) a new substance. The ash (can/cannot) be turned back into wood. Burning wood is a (physical/chemical) change.

   When lead is heated it (burns/melts). This is a (physical/chemical) change. More heating turns the lead into a (black/yellow/blue-black) powder or ash. This powder (is/is not) like lead. If it is a new substance the change is a (chemical/physical) change. We heated five metals. The one which burns with a bright white flame is (copper/lead/platinum/magnesium/iron).

   ### List of metals
   Aluminium
   Barium
   Calcium
   Copper
   Chromium
   Gold
   Iron
   Lead
   Magnesium
   Mercury
   Platinum
   Silver
   Tin
   Zinc   and others

Try to find one use for each metal in the list.

Burning

Heating substances can give new ones. When heated, some substances burn with a flame. Burning is a chemical change. It happens every day. We need to know more about it.

What do we know about it now?
In open air magnesium burns with a flame. So does wood. We heated wood inside a test tube. There was no flame.

Why is this?
Is it because nothing burns inside a test tube?
Try it and see! A gas jar is easier to use than a test tube.

Experiment 2.1 To burn a candle inside a gas jar

Put a candle on a gas jar cover. Light it. Invert a gas jar over it. The jar is full of air.

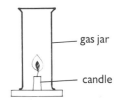
gas jar

candle

> What happens to the candle flame?
> Can air get into or out of the jar?
> We know what happens to the candle. What happens to the air?

Experiment 2.2   To find out what happens to air during burning

Light a candle. Stand it on a big cork. Float it on water in a trough. Put a bell jar over it. Stopper the jar.

The flame goes out. Light a splint. Hold it near the stopper. Take the stopper out. Put the flame in the jar.

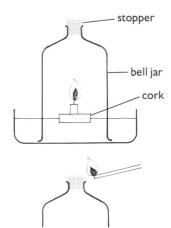
stopper

bell jar

cork

The jars in these two experiments are full of air. In both the flame goes out. No air can get in or out. Water rises into the bell jar. This means some air has gone. The air left in the jar puts out a flame. Air has two parts. One lets things burn. The other does not.

Why

In Science we try to explain why. The explanation is called a theory. It must explain all the facts. A fact is something we know to be true. We make theories all the time. Each theory uses the facts known at the time.

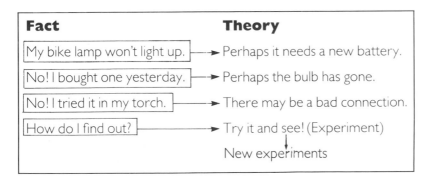

Fact	Theory
My bike lamp won't light up.	→ Perhaps it needs a new battery.
No! I bought one yesterday.	→ Perhaps the bulb has gone.
No! I tried it in my torch.	→ There may be a bad connection.
How do I find out?	→ Try it and see! (Experiment)
	New experiments

A theory of burning

A good theory explains all the facts we know. It can also show us how to find new ones. It suggests experiments to do.

The bell jar experiment gives us facts about burning. Do all burning substances give the same results?

Magnesium burns well. With care we can use it in a bell jar. Hook a spiral of it on a wire. Stick the wire into the stopper. Light the magnesium. Put it quickly into the bell jar. At the same time put in the stopper. The results are the same as before.

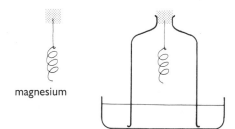

magnesium

Fact	**Theory**
Burning magnesium goes out in the bell jar. Water rises into the jar. Why?	If water rises, it is taking the place of air. Part of the air has been used.
The air left in the bell jar puts out a flame. Why is it extinguished?	The rest of the air does not let a flame burn in it. Air must be a mixture.
Magnesium turns into a white ash. Air is used up. Where is this air now?	The used-up air may be part of the white ash. So may the magnesium.

Full theory

Air is a mixture. Burning substances use up part of it. The rest of the air will not let flames burn in it. The used-up air may be part of the substance formed.

Magnesium burning gives a white ash. The theory says that

the white ash = magnesium + used-up air

Suppose the theory is correct. Then the white ash is two substances. It will have more matter than the magnesium. We can find out if this is true. Finding out will test the theory.

How do we measure matter?

Matter, mass and weight

The world is made of matter. The amount of matter in a substance is called its **mass**. Mass is measured in kilograms (kg). 1 kg = 1000 grams.

Hold 1 kg of iron. Let go of it. It falls towards the earth. The pull of the earth on it is its **weight**.

Put 1 kg on a spring balance. The pan goes down. Put 2 kg on the pan. It goes down twice as far. A balance can be used to measure mass or weight.

A top pan balance is best. It shows the mass on a moving scale.

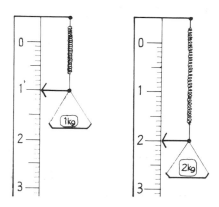

What happens to the air in burning?

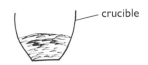
crucible

Experiment 2.3 To compare the mass of the white ash and the magnesium

Half fill a crucible with magnesium turnings. Put on its lid. Put it on a top pan balance. Read off its mass.

Put it on a pipe clay triangle, on a tripod. Heat it gently, then strongly.
Lift the lid with tongs regularly. Try to prevent any smoke escaping.
Let it cool. Put it back on the balance. Read off its mass.

before heating

Is white ash formed in the crucible?
Does the mass change?
If so, is it smaller or larger?

The magnesium turns into white ash. The lid is lifted to let air in.
It must be put back quickly. If not, white ash escapes as smoke.
The ash has a bigger mass than the magnesium.
It seems that

white ash = magnesium + used-up air

We have heated other metals. Some, like platinum, show no change.
Copper, lead and others give coloured powders.

after heating

Experiment 2.4 To find out if other metals show a change in mass

Put a strip of copper foil on a balance. Read off its mass. Hold it in tongs.
Heat it in the bunsen flame. Let it cool. Measure its mass.

What happens to the copper?
Is the mass the same, smaller or larger?
Does heated copper use up part of the air?
Is this air part of the black ash formed?

Check all the results in the class. Human beings make mistakes. Apparatus can fail. Other errors can happen. So a single result can be wrong. Several results give a more certain answer.

Black ash forms on every copper strip. It flakes off easily. This makes the mass less. Nearly all the strips show an increase in mass. The ash has a bigger mass than the copper.

Heated copper uses up air.

But does air have a mass? It seems not to fall to the earth.
How can we find out?
We can compare an empty flask with a flask full of air.

Willie's brother licks his chops,
Eats and eats and rarely stops.
As he gobbles up more matter,
George, of course, becomes much
  fatter.
Because he eats so much, alas,
His weight gets bigger,
  and his mass.

36

A top pan balance.

Questions

1. I burn magnesium in a closed space. When it goes out I put in a lighted splint. Explain how this shows that air is a mixture of gases.

2. Metals heated in air often leave an ash. Is the mass of the ash greater than the mass of the metal? Describe how you would find out.

3. Charcoal glows red hot when heated in air. Almost nothing is left. Where is the new substance if no ash remains?

4. Air is pumped out of a flask. Its mass is found to be 150 grams. Air is let in. The mass is found to be 151.2 g. What is the mass of the air in the flask?

5. Fill in the blanks.
   A piece of lead is put on a balance. This measures its _____. The lead is heated for a long time. The ash formed is _____ in colour. Its mass is _____ than the mass of the lead. The lead has used up part of the _____. The yellow ash contains _____ and _____ air. It is a _____ substance. Heating gives a _____ change.

Experiment 2.5 To find out if air has mass

Fit a round-bottomed flask with a stopper. The stopper carries a piece of glass tubing. On it there is a rubber tube and clip.

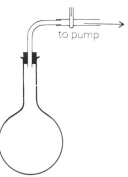

Connect the rubber tube to a pump. Pump out the air. Close the clip on the rubber tube. Put the flask on a balance. Read off its mass. Open the clip. Find out if the mass changes.

> What do you observe when the clip is opened?
> After opening, is the mass greater or smaller?
> What is the mass of the air in the flask?

The pump takes air out of the flask. Opening the clip gives a rushing noise. This is air rushing back into the flask. As a result the mass becomes greater. The increase is the mass of the air. It is very small.

Summary

Burning uses up air. Only part of the air is used. The rest of the air will not allow burning. Air is a mixture of gases. One part lets things burn. One part does not.

Burning forms new substances. They have a larger mass than the substance burnt.
This is true of magnesium which burns with a flame. It is also true of copper which does not. The used-up air is part of the new substance.
   the black ash = copper + used-up air

The mass of a substance is the amount of matter in it. It is measured in grams (g) and kilograms (kg). Mass is measured on a balance.

37

Oxygen and burning

Copper heated in air turns black. Our theory to explain this is:

copper + used-up air → black powder

The metal mercury is a silvery liquid. If it is heated a red powder forms on it. This is a chemical change. It is very slow. It may take days of heating to complete it.

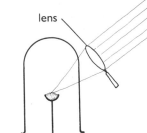
lens

In 1774, John Dalton heated this red powder. He had no bunsen burner. It was not invented until 1855. He put the red powder in a glass vessel. He used a lens to focus the rays of the sun on the powder. We shall get the same result with a bunsen burner.

The apparatus which Lavoisier used.

*Experiment 2.6 To heat the red powder

Do the heating in a fume cupboard. Put the red powder in a test tube. Heat it gently. Light a splint. Hold the flame at the mouth of the tube. Blow out the flame. Put the glowing red end of the splint inside the test tube. Let the tube cool. Tap its contents into a dish.

What happens to the flame of the splint?
What happens to the glowing splint?
Did you see anything coming out of the tube?
What is there in the test tube and dish?

A silvery metal forms. Drops of mercury fall into the dish. Mercury vapour is poisonous. This is why a fume cupboard is used.

The lighted splint burns more brightly. The glowing splint bursts into flame. A substance coming out of the tube causes this. But we saw nothing. This substance must be a gas.

The red powder splits up into mercury and a gas. This gas makes flames burn brighter. **It relights a glowing splint**.

Dalton told a French scientist about the gas. His name was Lavoisier. He saw a way of testing our theory of burning.

He used the apparatus shown. The vessel with a curved neck is a retort. It had some mercury in it. Lavoisier heated it on a charcoal fire.

After some days a red powder formed in the retort. The liquid rose inside the bell jar.

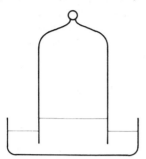

Some of the air had been used up by the mercury. Lavoisier took out the retort. He took every bit of red powder out of it. He heated the powder. It gave off a gas. All the gas was put back into the bell jar. The level of the liquid fell. It went back to the level at the start.

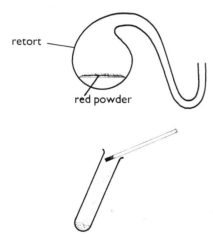

The gas put back was the used-up air. Lavoisier called it **oxygen**.

Now we know that:

Air is a mixture of gases. One gas allows substances to burn in it. This gas is oxygen.

The rest of the air is not oxygen. It does not let things burn in it. Most of it is a gas called nitrogen.

Burning substances use up oxygen. So do some heated metals. When all the oxygen has been used, burning stops.

metal + oxygen from air → a powder or ash

Elements and compounds

Heated mercury uses oxygen. It forms a red powder. In the red powder mercury and oxygen are joined together. We say that mercury **combines** with oxygen.

Heat splits the red powder up. Oxygen comes out of the tube. Mercury is left in it. The powder **decomposes**. This kind of chemical change is called **decomposition**.

Mercury cannot be split up. No one has been able to decompose it. Substances like this are called **elements**.

An element is a single substance. It cannot be split up into simpler substances.

Copper is an element. So is magnesium. Oxygen is an element too.

Elements combine. The result is called a **compound**. Magnesium combines with oxygen. The white powder formed is a compound. A compound contains two, or more, elements chemically combined.

The names of compounds

The name must say which elements it contains.
In the white powder are magnesium and oxygen.
Its name is **magnesium oxide**.
If it contains only two elements the name ends in **-ide**.
Why not call it oxygen magneside? Try saying it!

magnesium + oxygen
↘   ↙
magnesium oxide

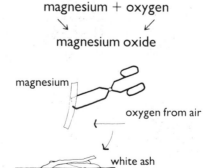

How much of the air is oxygen?

To find the answer we must
- measure some air
- take away its oxygen
- make sure all the oxygen has gone
- measure how much air is left.

Try to plan an experiment to do this.
How can we measure some air?
But we cannot see air.
How can we take away its oxygen?
What substances combine with oxygen?
As oxygen is used, more air will get in.
Then how shall we heat the metal?
How can we measure how much air is left?

Give it a title. Call it Experiment 2.7.
The best way is to find its volume.
Then use a vessel with volume markings.

Heat or burn a metal in it.
Then keep the vessel closed.
Put it in another vessel.
A syringe solves all these problems.

This is a syringe. It has air in it. Markings on the side tell us how much.

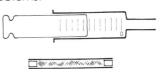

This is a glass tube. It is full of tiny bits of copper. Their colour is salmon pink.

Join the syringe and tube with rubber tubing. Join an empty syringe at the other end. Small bits of glass rod hold the copper in place.

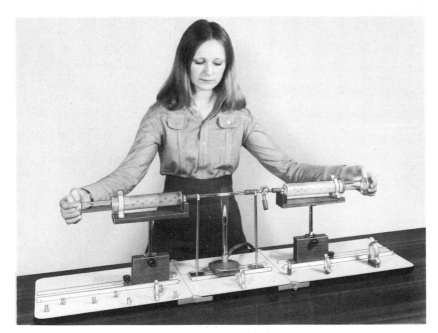

Two gas syringes.

Experiment 2.7   How much of the air is oxygen?

1. Heat a small length of the copper. Push the air out of the syringe. It goes through the copper. It enters the empty syringe. Heat the next section of copper. Push the air back through it.

Repeat this. Stop when there is no change in the volume of air. Let the tube cool. Note the volume of air left in the syringe.

> What happens to the heated copper?
> Does the last section of heated copper change?
> What does this tell us about the air?

The copper turns black. It takes oxygen from the air. So the volume of air gets less. The last section of heated copper stays pink. There is no oxygen left in the syringe.

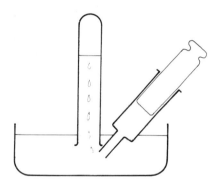

2. Push the gas out of the syringe. Collect it in a test tube. Test it with a lighted splint. Does the flame go out?

Here is a set of results: (the volumes are in cm^3)
Volume of air at the start = 41.5
Volume (cold) at the end = 33.0
Volume of oxygen used up = 8.5
41.5 cm^3 of air contain 8.5 cm^3 of oxygen.

The percentage of oxygen is $\dfrac{8.5 \times 100}{41.5} = 21\%$

Roughly $\frac{1}{5}$ of the air is oxygen. About $\frac{4}{5}$ is nitrogen.

Summary

When heated, mercury turns into a red ash. It combines with, and uses up, part of the air. Heating the red ash gives back a gas. Lavoisier proved that this gas is the used-up air. It is called oxygen. Substances burn more brightly in it than they burn in air. It makes a glowing splint burst into flame. Heated copper also takes oxygen out of air. We used this to show that 21% of air is oxygen. Air is roughly $\frac{1}{5}$ oxygen and $\frac{4}{5}$ nitrogen. Nitrogen puts out almost all burning substances.

Nitrogen, oxygen, copper, mercury are elements. They cannot be split up into other substances. Copper and oxygen combine. The black ash formed is not an element. It can be split up into copper and oxygen. It is a compound.

The compound is called copper oxide. The name ends in -ide. This shows that it contains only two elements. The name also shows which ones.

Questions

1. Copy this into your book. The brackets contain answers. Choose the *best* one. Write only this one into your book. Do not copy in the others.

   Magnesium is (an element/a compound/a mixture). It burns giving a (black/white/red) ash. It combines with (nitrogen/oxygen/gas) in the air. The ash is (an element/a compound/a mixture). It is called (oxygen magnide/magnesium oxide).

   Heated mercury gives a (black/red/white) ash. This is a (physical/chemical/colour) change. The change is (very fast/slow/very slow). The ash is called (mercury oxide/oxygen mercuride). When heated the red ash splits up. It gives a gas called (oxygen/nitrogen/mercury). The split-up is a (physical/chemical/colour) change. It is called (combination/decomposition).

   Air is a (compound/mixture) of several gases. Oxygen is about ($\frac{1}{5}$/$\frac{4}{5}$/ 79%) of the air. Substances burn in oxygen (feebly/very well). Oxygen (puts out/relights) a glowing splint.

2. Make a list of all the substances we burn. Do they use oxygen from the air? Why is the air never short of oxygen?

3. Burning substances need oxygen from the air. Explain how a fire blanket works.

Pure oxygen

Oxygen is made in two ways. It can come from oxides or from air.

Air is a mixture of gases. To get oxygen the other gases must be taken away. Oxygen is manufactured by this method.

Oxides are compounds. Heat decomposes some of them giving oxygen. Mercury oxide is one of these. But it is expensive. Hydrogen peroxide is cheap and easy to use.

*Experiment 2.8 To make and collect oxygen

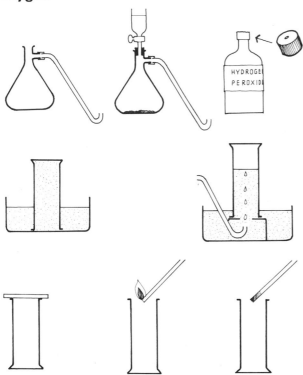

The drawing shows a conical flask (shaped like a cone). It has a side neck. Connect a glass tube to it. Put black manganese (IV) oxide into the flask. Fit a dropping funnel into the stopper. Fill this with hydrogen peroxide solution.

Fill a trough with water. Fill a gas jar with water. Turn it upside down in the trough.

Open the tap in the funnel. Let some hydrogen peroxide drop into the flask. Close the tap.

Gas comes out of the glass tube. Stand the gas jar on a shelf. Put the tube under it.

When the jar is full of gas, slide a cover over it. Take it out. Fill six more gas jars.

Use one jar to find out if the gas is oxygen. Put in a lighted splint. Blow the flame out. Put the glowing splint in the gas.

We shall burn elements in oxygen. Oxides will be formed. We shall try to dissolve each oxide in water. We shall test the water. The tests are litmus paper and Universal Indicator.

On page 22 we dissolved the green substance out of grass. Litmus is made in the same way. It is the dye in one kind of moss. The dye is mauve in colour (like pickling cabbage). Filter paper is soaked in the solution of it. Drying it gives litmus paper.

Universal Indicator is a solution. It contains coloured dyes.

They are both called **indicators**. They indicate, or show, what a liquid contains.

Burning in pure oxygen

Hydrogen peroxide splits up on its own. It gives oxygen very slowly.
This explains the hole in the bottle cap. It lets this oxygen escape.

Manganese (IV) oxide speeds up this change. It does so by being there.
It is still there at the end. It is not changed by the reaction. Substances
which do this are called **catalysts**.

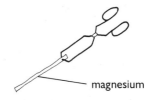
magnesium

*Magnesium   Hold a strip of magnesium in tongs. Light it at the bunsen. Hold it in a gas jar of oxygen. *Do not look at it.* Watch it only through blue glass.

In each test: Look at the results in the gas jar. Add water. Put a cover on the jar. Shake it. Put in a litmus paper. Add drops of Universal Indicator.

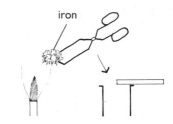
iron

*Calcium   Hold a piece of calcium in tongs. Burn it in the same way. *Do not look at the flame.*

*Iron   Put a bunsen flame near the top of a gas jar. Hold a tuft of iron wool in tongs. Touch the flame with it. Put it quickly into the oxygen.

*Carbon   One form of carbon is charcoal. Hold a stick of it in the flame. Put the red-hot end into oxygen.

*Sulphur   Put sulphur in a combustion spoon. Hold it in the bunsen flame. When it burns, hold it in oxygen.

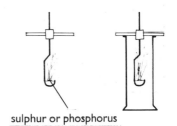

*Phosphorus   Put red phosphorus in another spoon. Get it burning gently. Put the spoon into a gas jar of oxygen.

sulphur or phosphorus

Have this table ready to put in the results.

element	metal or non-metal	colour of flame	oxide is	dissolves in water	litmus colour	Univ. Ind. colour
calcium	metal	brick red	solid	perhaps	blue	violet
carbon	non-metal	white	gas	perhaps	mauve	yellow

All these elements burn brilliantly in oxygen. Some of the oxides are
solid. Others cannot be seen. They are likely to be gases.
In water litmus is mauve; Universal Indicator is green.

Water is added to each gas jar. The indicator colour may change.
Water alone cannot do this. It must have reacted with the oxide.
A new substance is formed. It affects the indicators.

Acid and alkali

The burning elements form oxides. Some of these dissolve in water. The oxide in water gives a solution. Most solutions make litmus red or blue. They also change the colour of Universal Indicator.

Experiment 2.9 To find out what substances affect indicators

Put vinegar in two test tubes. Put lemon juice in two others. Add drops of litmus to one of each. Add drops of Universal Indicator to the other two.

Try the taste of vinegar and lemon juice. (Yes, it is safe.) What effect do they have on the indicators?

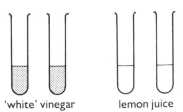

'white' vinegar   lemon juice

Vinegar and lemon juice taste sour, sharp or acid. They are acidic solutions. Each contains a substance called an acid.

Put sodium hydroxide solution in two test tubes. Test each of them with one of the indicators.

Sodium hydroxide solution is alkaline. It contains a substance called an **alkali.** The family name of alkalis is **hydroxide.**

sodium hydroxide solution

Indicators

Acid turns litmus red. Alkalis turn litmus blue. Many dyes change colour in this way. They are called indicators. They indicate, or show, that acids or alkalis are present.

Universal Indicator is a mixture of dyes. It shows many colours. There are usually seven. They are the colours of the rainbow.

Red   Orange   Yellow   Green   Blue   Dark Blue   Violet

acid colours   alkali colours
stronger   weaker   weaker   stronger

Red shows a stronger acid than orange or yellow. Violet shows a stronger alkali than blue or dark blue.

Alkalis

In Experiment 2.8 we burnt elements in oxygen. Check the metals in the table of results. They were magnesium, calcium and iron.

All three gave oxides. Two of the oxides dissolve in water.

> *Calcium burns in a blinding flash*
> *And changes to a dull white ash.*
> *In the dull white ash I find*
> *Metal and oxygen combined.*
> *So, bless my cotton socks, I'd*
> *Better call it calcium oxide.*

What indicator colours did the solutions give? Are they acid or alkaline solutions?

Take calcium, for example. Its oxide dissolves in water. The solution turns litmus blue. It has an alkali in it. The name given to an alkali is hydroxide. This one is calcium hydroxide.

Check the Universal Indicator colour it gives. Is the alkali strong or weak?

We can write in short:

calcium + oxygen → calcium oxide
calcium oxide + water → calcium hydroxide

Now check the results for a non-metal. Try carbon.

Carbon burns with a white flame. We cannot see the oxide. It must be a gas. It is called carbon dioxide.

Carbon dioxide dissolves in water. We know this because the water in the gas jar turns Universal Indicator yellow. This shows it is a weak acid. It is called **carbonic** acid. The name shows which element it came from.

We can write in short

carbon + oxygen → carbon dioxide
carbon dioxide + water → carbonic acid

Half-way or neutral

The half-way colour for litmus is mauve. It will be mauve if there is no acid or alkali. The liquid is **neutral**.

Carbon in oxygen burns white hot.
What kind of oxide have we got?
Nobody saw one in my class.
Carbon dioxide must be a gas.
It can't be seen but it's there all right.
It turns lime water from clear to white.

What will happen if we mix acid and alkali?

Experiment 2.10 To mix an acid and alkali

Burn a stick of charcoal in oxygen. Shake the oxide with water.

Burn calcium in air or in oxygen. Shake the oxide with water.
Filter the mixture. Add litmus paper to the clear solution.
Pour it into the carbon dioxide. Put black paper behind the jar.

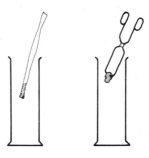

What is formed when carbon burns?
What do you see in the other gas jar?
What happens to the litmus?
What happens when the two solutions mix?

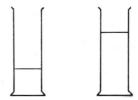

Carbon burns to give carbon dioxide.
Calcium gives calcium oxide. This reacts with water. The litmus turns blue. The clear solution is an alkali. It is calcium hydroxide solution. Carbon dioxide turns it 'milky'. The black paper helps to show this.

Mixing *Acid* and *Alkali*

Pure water has no acid in it. It is not acidic. It has no alkali in it. It is not alkaline. In pure water litmus is mauve. This colour is half-way between red and blue. Litmus shows that water is **neutral**. The half-way colour of Universal Indicator is green. Green shows neutral.

Carbon dioxide in water forms an acid. Calcium oxide in water forms an alkali. Mixing the two gives a white substance.

Calcium hydroxide solution is 'lime water'. Carbon dioxide makes it 'milky'. Only carbon dioxide can do this. We cannot see carbon dioxide. We can recognise it because

> Carbon dioxide makes lime water milky.

Experiment 2.11 To mix another acid and alkali

Use acid and alkali from the bench bottles. One-third fill a test tube with sodium hydroxide solution. Add drops of Universal Indicator.

Fill a dropper with dilute hydrochloric acid. Add drops of it to the alkali. Shake the test tube. Add more acid. Stop adding it when the indicator changes colour.

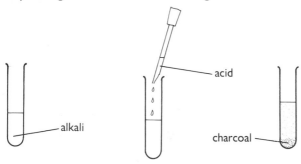

Add a pinch of charcoal to the test tube. Heat it. Filter the liquid into a dish. Heat the dish. Stop when most of the water has gone. Put some of the liquid on a glass slide. Look at it through a microscope.

new substance

Pure salt crystals.

What colour is the indicator in alkali?
How many indicator colours did you see?
Was the final colour green?
What does charcoal do to the solution?
Are crystals formed on the glass slide?
Do their shapes show what the solid may be?

You may have got all the colours from red to violet. It is difficult to get a neutral solution. We shall try again later with another method.

Charcoal takes out the colour. Heating the solution gives a white solid. It may appear as crystals. Are they like those in the photograph?

acid + alkali → a new substance (salt?)

Calcium + oxygen → calcium oxide is a summary. It means 'Calcium combines with oxygen. Calcium oxide is formed.' We can write these **word equations** for any reaction.

Iron + oxygen → magnetic iron oxide
This oxide does not react with water.

Sulphur + oxygen → sulphur dioxide
Sulphur dioxide + water → sulphurous acid

Some metals have more than one oxide. Copper has two. Their names must show which one we mean. We have used **black** copper oxide. There is a better naming method. We shall learn it later. It uses Roman numbers; I, II, III, IV and so on.

Phosphorus + oxygen → phosphorus (v) oxide
Phosphorus (v) oxide + water → phosphoric acid

Summary

Oxygen is made from air or from oxides. The best oxide for lab use is hydrogen peroxide. Manganese (IV) oxide is a catalyst. It makes hydrogen peroxide decompose quickly. The gas is collected 'over water'. We know the gas is oxygen. It relights a glowing splint.

Elements burn more brightly in oxygen than in air. Oxides are formed. Some metal oxides dissolve in water. Each solution turns litmus blue. It is alkaline. It contains an alkali.

Non-metal oxides may dissolve in water. These solutions are acidic. Each contains an acid. Acids turn litmus red. Universal Indicator is a mixture of dyes. It shows the colours of the rainbow. They run from red to violet. Red shows acid, violet is alkaline, green shows neutral.

A word equation sums up a chemical change. phosphorus + oxygen → phosphorus(v) oxide Acid plus alkali can give a neutral solution. A new substance is formed. In this way, carbon dioxide makes lime water 'milky', or cloudy.

Questions

1. Sodium is a metal. It burns well in air. Choose the best answer in each statement below.

   a) In oxygen sodium burns (brilliantly/well/fairly well/feebly/not at all).
   b) Sodium oxide dissolves in water. The solution turns litmus (red/mauve/blue/violet).
   c) This solution is (acidic/alkaline/neutral).
   d) In this solution, Universal Indicator becomes (red/orange/yellow/green/violet).

2. Explain why any substance burns more brightly in oxygen than in air.

3. You are given an element. How would you find out if it is a metal or a non-metal?

Things to do

Use vinegar as an acid. Use washing soda as an alkali. Find out if any of these are indicators: beetroot juice, blackcurrant juice, elderberry juice. Use one of them to test Milk of Magnesia, tap water, Alka-seltzer.

Hydrogen peroxide gives oxygen. Manganese(IV) oxide speeds up the change. Find out if the change is speeded up by (a) black copper oxide, (b) magnetic iron oxide, (c) blood.

*Acid gives red,
Alkali blue
If you use litmus
This rhyme is true*

Other burning substances

Burning is very common in everyday life.

We burn things by accident. Houses, forests, oil tankers and clothing are all set on fire by chance.

Stubble is straw left after harvesting corn. Burning it kills plant pests. It kills useful creatures too. It spreads smoke and soot. It can get out of control. It wastes a useful fuel.

We burn some substances to keep ourselves warm. Coal, coke, gas, wood and oil are used in this way. They are called fuels. They give heat energy. It keeps houses and other buildings warm.

We need transport. Cars, buses, lorries, ships, trains all burn fuel.

We burn fuel to cook food. We may use electricity for cooking and transport. However, fuels are used in power stations to make electricity.

When fuels burn they form oxides. These oxides are called the 'products of combustion'. When magnesium burns we can see the oxide. It is a white powder. When fuels burn we do not often see the oxides. This means that they are gases or vapours.

Smoke comes from not burning a fuel completely. It wastes fuel and pollutes the air.

A candle burns in air. What is 'candle oxide'?
Petrol burns. What is 'petrol oxide'?

Burning stubble in a wheatfield.

Coal arriving at a power station.

A bush fire in Australia.

The products of combustion

Experiment 2.12 To find out what these products are

Stand a candle on a tile. Hold a funnel over it. Connect the funnel to the test
tubes. Connect a filter pump to the other end of the apparatus.

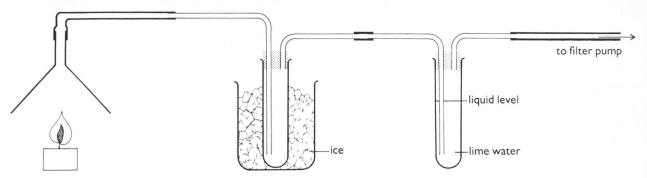

The first test tube is empty. It has ice round it.
The second test tube is half filled with lime water.
Light the candle. Turn on the filter pump.
Use other fuels instead of a candle.

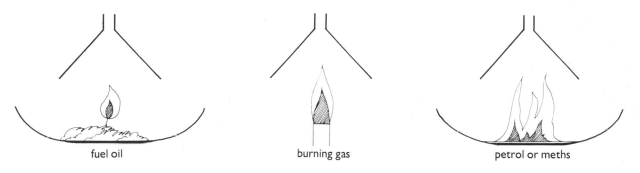

| fuel oil | burning gas | petrol or meths |

The pump draws air through the apparatus. The air takes the products of
combustion with it. They pass through the tubes.

   What do you see in the first test tube?
   Can you guess what this substance might be?
   What happens to the lime water?

Remember that most products of combustion are gases or vapours. The lime
water 'turns milky'. This happens with all the fuels we burn. One of the
products of combustion is carbon dioxide.

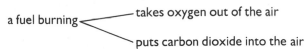

a fuel burning ——— takes oxygen out of the air

——— puts carbon dioxide into the air

The first test tube is cold. It will cool any vapours. It may turn some of them
back into liquids. The liquid which collects in this tube is colourless. It looks
like water.

   Fuel + oxygen → carbon dioxide + liquid

49

Is water formed when fuels burn?

All burning fuels give carbon dioxide. Many also give a liquid. This has no smell or colour. It looks like water. How can we find out if it is water?

> What properties of water do we know?
> What is its boiling point?

There are two more tests we can use.

Experiment 2.13 To heat copper sulphate crystals

Put the blue crystals into a test tube. Fill it to a depth of 1 cm. Heat it gently. Collect any liquid in a cooled, dry test tube. Boil this liquid. Record its boiling point. Let the solid get quite cold. Then add water.

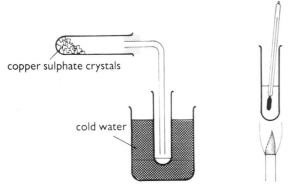

copper sulphate crystals

cold water

> What happens to the heated blue crystals?
> Does boiling point show the liquid is water?
> What does water do to the white substance?
> Do the water and solid become hot?

Experiment 2.14 To heat cobalt chloride crystals

Heat cobalt chloride crystals as in Experiment 2.13.

The two results show one difference. What is it?

The blue crystals lose a liquid. They turn into a white solid. The liquid boils at 100°C. This shows that it is probably water. Then only water can reverse the change. The water we added did this. The white solid is called anhydrous copper sulphate. It can be a test for water.

Cobalt chloride crystals lose water. A blue solid forms. It is anhydrous cobalt chloride. Only water

Traffic burns fuel.

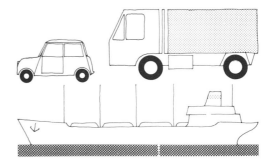

can make it pink again. Like copper sulphate, it shows water is there. It does not show that the water is pure. Boiling point does that. Anhydrous means 'without water'.

Take the liquid from burning fuels. It boils at 100°C. Add white anhydrous copper sulphate. It turns blue.

Water is a product of combustion of fuels.

Soak filter paper in pink cobalt chloride solution. Dry it. Hold it high above a bunsen flame. It turns blue. Cut it into strips. Use them as test papers for water. Water turns them pink again.

Fuels and the air

Burning fuels use oxygen. They take it from air. One product is carbon dioxide. Another is water. We all burn fuels. It happens all over the world. In Britain alone there are over 17 million cars and lorries. They burn petrol or diesel oil. The fuel vapour is mixed with air. The mixture is fed into the engine.

It burns in the cylinders. The products go through exhaust pipes into the air.

Every fuel contains carbon compounds. Most of it burns to carbon dioxide. Not all of it does. The mixture contains too little air. Some fuels form carbon monoxide. This gas is poisonous. Carbon itself may appear as smoke. Other exhaust gases are also formed. They all cause air pollution.

The picture shows a traffic hold-up. Cars are not moving. Some engines are still running. Exhaust gases pour out. Carbon monoxide may build up to high levels. In still air this can be dangerous.

Trains and tractors use diesel oil. Jet aircraft use kerosene. Ships burn heavier oils as fuel. So does oil-fired central heating. We also use solid fuels. The common ones are wood, coal and coke.

There are many gas fuels. They have advantages. Solid and liquid fuels need transport. Gas can be sent through pipes. This has dangers. Pipes can rust or crack. A gas leak mixes gas with air. A spark or flame can explode it. Damage can be huge.

Natural gas is the main gas fuel (see page 202). Others are coal gas, propane, butane and Calor gas.

Burning fuel takes out oxygen

↑

the air

↑

Burning fuel puts in carbon dioxide and water

This raises three questions:
   Why has all the oxygen in air not been used up?
   Why is the air not full of carbon dioxide?
   Is there any carbon dioxide in the air at all?

Summary

Human beings find it useful to burn fuels. They keep us warm. They allow us to cook food. They provide energy to run our transport. They burn in power stations making electricity. They put carbon dioxide and water into the air.

Water can be recognised by its boiling point. Other tests are with white anhydrous copper sulphate and blue anhydrous cobalt chloride. They prove that water is present. They do not prove that it is pure. Boiling point does that.

Burning fuels can pollute the air. Without enough oxygen they form smoke and carbon monoxide. They use up the oxygen of the air. They put into air carbon dioxide and water vapour.

Questions

1. Give three reasons why we burn fuels.

2. Burning fuels affect the air around us. What changes do they cause? How do they cause these changes to happen?

3. A candle burns in air. Draw the apparatus used to collect the products of combustion. Label the drawing.

4. I have a colourless liquid. How would you prove that it contains water? What test will show that it is pure water?

5. What do we mean by air pollution?

6. It is a cold wet evening. You have just come home. You can choose to light a coal fire or a gas fire. Which will you choose? What are the advantages of each? Do they have disadvantages? If so, what are they? Give reasons for your choice.

Carbon dioxide and the air

*Experiment 2.15 To find out if air contains carbon dioxide

  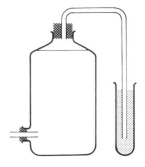

This is a large jar full of air. Connect it to the water tap.

Run water into the jar. This will push air out at the top.

Let this air slowly bubble through lime water.

We are using the kind of air we breathe in.

Use the same amount of lime water again. Take a deep breath. Blow gently into the lime water.

We are using the kind of air we breathe out.

> What happens to the lime water?
> In which test tube was it more milky?
> Which air has more carbon dioxide?

A whole jar of air made the lime water just cloudy. The amount of carbon dioxide in it must be small. Very little breathed-out air was needed. The lime water was very milky. This air has far more carbon dioxide.

There are about 2500 million people on earth. Nobody knows how many animals there are. They all breathe in air. This air has little carbon dioxide in it.

They breathe out. This air contains much more carbon dioxide. We all put carbon dioxide into the air when we breathe out.

52

What about a bite to eat?

What is your temperature?
Are you hotter than the air around you?
Does your body lose heat to the air?

37.3 °C   20 °C or less

Do you run? Do you ride a bike? Do you play games? Do you stand still? To do all these needs energy. Where does it come from?

*Experiment 2.16 To burn a cornflake

Hold a large cornflake in tongs. Light it at the bunsen flame. Hold it in a gas jar of oxygen. Let it burn. Shake lime water in the jar.

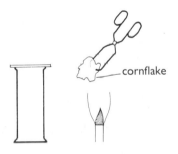

cornflake

What does the lime water prove?

Most foods behave like the cornflake. They burn. This means that they give heat. They also give carbon dioxide.

Food is digested inside us. Chemical changes turn it into substances we can use. These are carried by the bloodstream.

We breathe air into our lungs. Oxygen from it passes into the blood. It is carried round the body. Food and oxygen meet in muscle tissue. The food 'burns', or **oxidizes**. It uses oxygen.

We use food as a fuel. Oxidizing it gives heat. This heat keeps us warm. It provides energy for movement. But burning, or **oxidation**, also forms carbon dioxide. This passes into the blood. It is carried back to the lungs. We breathe it out.

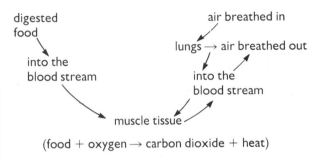

digested food

into the blood stream

air breathed in

lungs → air breathed out

into the blood stream

muscle tissue

(food + oxygen → carbon dioxide + heat)

Running is one of Willie's crazes.
It makes him get as hot as blazes.
What gives the heat when Willie hustles?
Food oxidizing in his muscles!

Summary

There is little carbon dioxide in the air. The air we breathe out contains much more. Food burns or oxidizes inside us. It uses oxygen from the air we breathe in. This produces carbon dioxide and heat. The heat keeps us warm. It provides energy for movement. The carbon dioxide passes into the blood stream. It is breathed out by the lungs.

We use oxygen from the air

The air round the earth is called the atmosphere. It is a mixture of several gases.

I have just breathed in. My lungs now contain 1000 cm³ of this air. The table shows the main gases in it. It shows their volumes, too.

Name of gas	The air I breathe in	The air I breathe out
oxygen	208	170
nitrogen	781	781
carbon dioxide	0.3	40
others	10	10

My blood takes oxygen from this air. It puts carbon dioxide into my lungs instead. I breathe out. The table shows the gases in this air, too.

Thousands of millions of animals live on Earth. They include us. We all breathe. We cannot live without oxygen. Look at the change we make in the atmosphere. Compare the two sets of figures.

The oxygen volume falls. It drops from 208 to 170 cm³. Carbon dioxide rises from 0.3 to 40 cm³. We know about burning fuels. They also take oxygen out of air. They put in carbon dioxide.

Yet the atmosphere shows only a slight change. The gas proportions stay roughly the same. So some other change must happen as well.

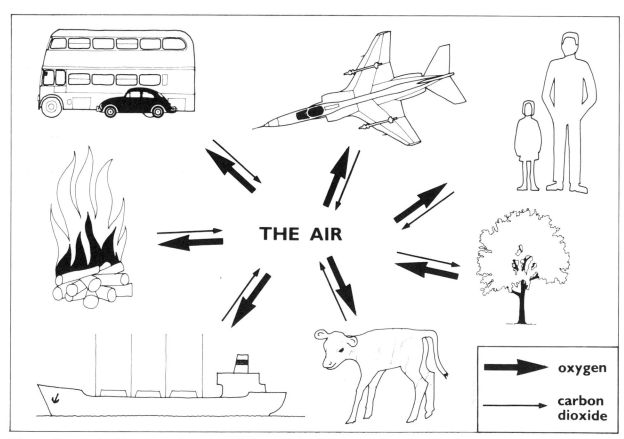

THE AIR

→ oxygen

→ carbon dioxide

The carbon cycle. Man, animals and machines all take oxygen out of the air.
Only green plants put it back in.

Green plants put back oxygen

*Experiment 2.17 Plants and the atmosphere

1. Put water weed under the funnel. Stand the trough in sunlight or daylight. Collect gas in the test tube. Test it with a lighted splint. Then test it with a glowing splint.

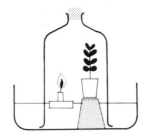

What gas does water weed produce?

2. Put the plant on a stand. Light the candle. Stand the bell jar over both. Stopper the jar. Leave the apparatus in the window in light.

Set up the apparatus a second time. Put it in a cupboard. Leave both for three days. Test the air in each with a lighted splint.

What do the lighted splints show?

The water weed gives a gas. It does this best in sunlight. This gas relights a glowing splint. It is oxygen. Some of it dissolves in the water. Most of it passes out into the air.

The candles go out. They have used up all the oxygen. The plants live in this air. In the daylight jar the splint stays alight. The plants must have given oxygen. In the other it goes out. No oxygen has been produced in the dark.

All green plants in light give oxygen.
This oxygen passes into the atmosphere.

Green plants
● take carbon dioxide from the air
● take in water through their roots
● use these substances to make starch or sugar.

This chemical change happens in the leaves. Green chlorophyll speeds it up. The energy needed comes from sunlight.

The change is called **photosynthesis**. It means the building up of carbon compounds using light. *Photo* means light and *synthesis* means build.
As well as starch, oxygen is formed. It passes into the air.

carbon dioxide + water + energy →
starch + oxygen

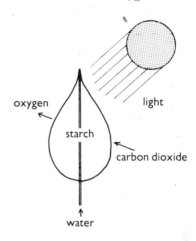

Animals oxidize their food. Fuels burn. Both changes use up oxygen and give carbon dioxide and energy.

Photosynthesis does the opposite. It just balances breathing and burning. This means the atmosphere is not changed very much.

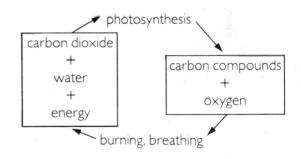

These changes are called the **carbon cycle**, or **energy cycle**.

The manufacture and uses of oxygen

Substances burn more brightly in oxygen than in air. We need oxygen to live. These two facts explain its uses.

A fuel gas, such as acetylene, burns in air. In oxygen its flame reaches 3000 °C. This melts metals: Melting a metal is one way to cut it. Liquid metal is used to join metals by welding.

Space rockets burn fuel. Tonnes of it burn each second. This needs huge amounts of oxygen. It is carried in the rocket as liquid. The gases formed drive the rocket upwards.

Iron is made in a blast furnace. Oxygen and air are blown in. White-hot liquid iron comes out. It is not pure. Oxygen is blown into it. Elements such as carbon and phosphorus burn away. The iron is then pure. It can be made into steel.

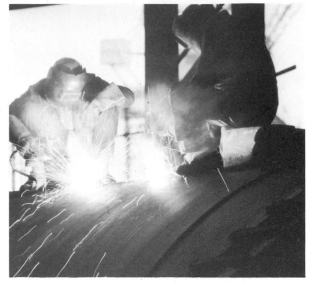

Joining metals by welding.

Tapping a blast furnace.

A baby in an incubator with an extra oxygen supply.

Putting hot iron into a vessel so that oxygen can be blown through it.

Breathing normal air gives us enough oxygen. Higher up the air gets thinner. It does not give enough oxygen. Mountain climbers and fliers take it with them, stored in cylinders.

People who are ill may be very weak. Breathing may tire them. They may have lung diseases. They may be under anaesthetic. Oxygen is fed into the masks through which they breathe.

Divers use compressed air to give oxygen. Its nitrogen dissolves in the blood. It comes out only slowly. It can give the diver 'the bends'. Using a mixture of oxygen and helium avoids this. It is used by climbers as well as divers.

The manufacture of oxygen

Manufacture means making large amounts. Oxygen is made from air. Water and carbon dioxide are taken out first. They are frozen to solids.

You have used a bicycle pump. Air compressed by the pump gets hot. Letting this air escape will cool it. A pump is used in making oxygen.

The pump gives hot compressed air. This passes into water-cooled pipes. It is now cold compressed air. It escapes through a valve. This cools it still more. It is used to cool the air coming down to the valve. This comes out even colder.

At last the air is cold enough to turn to liquid. This is liquid air. It is at about −200 °C. It is a mixture of liquid oxygen and nitrogen.

As it warms, nitrogen boils off. It boils at −196 °C. Liquid oxygen remains. Its boiling point is higher. It is −183 °C. (Higher ?) Oxygen is stored as a liquid. For most uses it is sold in steel cylinders, under pressure.

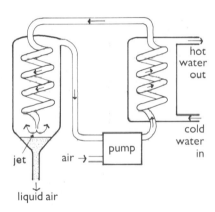

Copy this out. Fill in the blank spaces.

The air round the Earth is called its _____. It contains two main gases, _____ and _____. 78% of air is _____. 21% is _____. 0.03% of the atmosphere is *Ca*_____ _____.   *CO₂*

Oxygen is taken out of air in two main ways. It is used by animals _____. It is also used by fuels _____. Both changes also put the gas _____ _____ into the atmosphere.

Green plants take _____ _____ out of the air. Their roots take in _____. They use these two substances to make _____ or _____. This change also gives the gas _____. It happens in the _____ of the plant. This puts _____ into the air. The change is called p_o_o_y_t_e_i_. It means building up in _____.

Photosynthesis balances _____ and breathing. The gas proportions in air remain _____.

Oxygen has many uses. Gas fuels burn more _____ in it than in air. The flame is very _____. It can melt _____. It is used in cutting and _____ metals. Oxygen is used in space _____. It is carried as a _____. It makes fuel burn very _____. The gases formed _____ the rocket.

Questions

1. Give three reasons for using oxygen masks.

2. What is oxygen used for in making iron and steel?

3. Substances A, B, C, D and E are heated in oxygen.
   A burns to a white powder. This dissolves in water. The solution turns litmus blue.
   B burns giving carbon dioxide and water.
   C burns giving an oxide. This oxide is a gas. It dissolves in water. It turns litmus red.
   D shows no sign of change.
   E does not burn. A black ash forms on it.

   Which substance, A, B, C, D or E, is a metal? Which substance, A, B, C, D or E, is most like (a) copper, (b) salt, (c) platinum, (d) a candle, (e) a cornflake, (f) magnesium, (g) wood, (h) sulphur, (j) calcium, (k) charcoal, (m) petrol?

Water – is it an oxide?

Read these sentences. Write down the missing words.

A substance which burns uses _____ from the air. The new substance formed is called an _____. It is made up of the burning substance + _____. Most burning fuels give carbon _____ and _____. Water is formed by burning. So it may be an

_____.

Check your answers with those at the bottom of the page.

Magnesium is an element. It combines with oxygen when it burns. The white ash is magnesium oxide. In it two elements are combined. Any oxide contains oxygen and some other element.

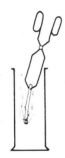

If water is an oxide it contains oxygen. This is combined with some other element. Which one? Take oxygen away from water. This will leave the other element. What substances take away oxygen? Those which burn well in air. Magnesium burns brilliantly.

water = X oxide
water − oxygen → X

Experiment 3.1 To try to get an element from water

Clean magnesium ribbon with fine sandpaper. Cut off a 3 cm length. Fold it loosely in three. Drop it into water in a test tube. Watch what happens. Now warm the test tube. If the metal moves, tap the tube gently.

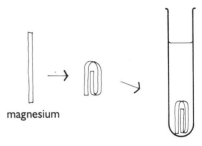

magnesium

What do you see happen in the test tube?
Can you see any signs of a gas forming?
Does the magnesium move in the water? Why?
Does the magnesium change in any way?

Magnesium is a normal metal. It has metal properties. It has a shiny surface. It will sink in water.

Sand paper cleans off any tarnish. In water, tiny bubbles form on the metal surface. They act like balloons. They carry the magnesium upwards. Tapping the tube shakes the bubbles off. The metal sinks.
The magnesium slowly changes. It loses its metallic lustre. Its surface becomes tarnished.

Magnesium is a good taker of oxygen. It could take oxygen from water. If it does, the tarnish could be magnesium oxide. The gas would be the element. Water would then be 'bubbles oxide'. Invent a way of collecting this gas. We need enough to test it.

Collecting the bubbles

The bubbles are a gas. They are very small. They are given off slowly. How can we get more gas more quickly?

Experiment 3.2 To collect the gas

Take all the bits of magnesium used by the class. Clean them with sandpaper. Put them in a trough. Put a funnel over them as shown.

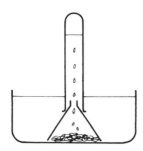

Fill a test tube with water. Invert it in the trough. Put it over the funnel. Let enough gas collect. Test it with a lighted splint.

> Do you see any new results?
> What happens to the lighted splint?

The gas still comes off at a slow rate. It takes some time to collect enough. The splint goes out. There is a small explosion. This is often called a 'squeaky pop'.

This test makes the gas easy to recognize.

What other metal burns brilliantly?
What is it which . . 'burns in a blinding flash
            And changes to a dull white ash'?

Experiment 3.3 To try to get the gas more quickly

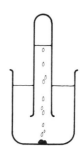

Two-thirds fill a small beaker with water. Invert a test tube full of water in it. Drop in a tiny piece of calcium. Hold the test tube over it. Collect the gas. Test it with a lighted splint. Put a litmus paper in the water.

> Does calcium react faster than magnesium?
> Does it give the same gas as magnesium?
> Litmus turns blue. What kind of substance is there in the water? What is its name?

Both metals take oxygen from water. Shiny magnesium becomes dull. We say it 'tarnishes'. The tarnish may be the oxide. The calcium reacts faster. The test tube soon fills with gas.

The water contains an alkali. The family name of alkalis is 'hydroxide'. The water is a solution of calcium hydroxide.

Not all the alkali dissolves. This is why the water is cloudy. Enough dissolves to turn the litmus blue.

The gas is **hydrogen**. A lighted splint gives a small explosion. Hydrogen is an element. **Water is hydrogen oxide**.

The reactants are calcium and water. The products are hydrogen and calcium hydroxide. Write an equation.

Hydrogen

Calcium + water →
calcium hydroxide + hydrogen

Water is hydrogen oxide. Hydrogen means 'water producer'!

From the last experiment we know some properties of hydrogen. It is a gas. It has no colour or smell. Bubbles of it rise through water. So it does not dissolve much in water. To discover other properties we need larger amounts. It is best made using metals and acids.

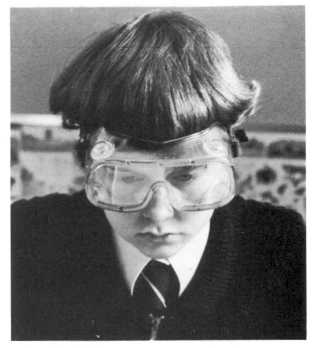

Remember to wear eyeshields.

We need to collect hydrogen in gas jars. So we choose a metal which gives hydrogen fairly quickly.

Experiment 3.4 The action of acids on metals

Put a strip of metal in a test tube. Add dilute hydrochloric acid. If bubbles appear, test them. Put a lighted splint at the mouth of the tube. Wear eye shields. Use a test tube holder.

Try this with copper, zinc, aluminium, lead, tin, iron and magnesium. There may be no reaction. If so, heat the tube gently. Do not let the liquid boil.

Write your results in the form of a table. Make a list of the metals. Put at the top the one which reacts best. Put the others in order of how quickly they give hydrogen.

Metal used	Is heat needed to get gas?	Is hydrogen formed?	How quickly is it formed?

*Experiment 3.5 Other properties of hydrogen

Put lumps of zinc in the flask. Add a few drops of copper sulphate solution. Pour dilute hydrochloric acid down the thistle funnel. Add enough to cover the bottom of the funnel.

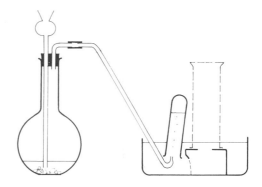

First collect the gas in test tubes. Test each tube with a lighted splint. Wait until the gas gives no 'squeaky pop'. Then collect it in gas jars over water. Why do we test it in this way?

*The properties of hydrogen

Put safety screens round the gas jars in tests 1 and 2.

1. Take a gas jar of hydrogen. Hold a lighted splint near the top. Quickly take off the lid.

2. Take the lid off a gas jar of hydrogen. Count ten. Put in a lighted splint.

3. Take a test tube half full of water. Collect hydrogen in it. Test it with a lighted splint.

4. Add detergent to water. Put in a little glycerol. Stir it. Dip the hydrogen delivery tube into it. Blow bubbles. Shake them off the tube. Chase them with a lighted splint.

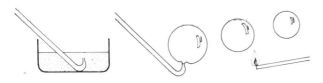

5. Fill a balloon with hydrogen. Use a cylinder of gas. Tie the balloon with string at the neck. Let the balloon go.

What do the tests tell us about hydrogen?

In 1. pure hydrogen burns quietly. In 2. the splint burns quietly. Counting ten gives time for the hydrogen to escape. In 3. it is mixed with air. The splint makes it explode.

The bubbles and the balloon are full of hydrogen. They rise into the air. Hydrogen is less dense than air. This is why it escapes from the open jar. Air takes its place.

An air-hydrogen mixture explodes. This is why we first collect it in test tubes. A gas jar of the mixture would be dangerous in test 1. When burning or exploding it forms an oxide.

*Experiment 3.6 To show that hydrogen oxide is water

Use the apparatus shown. Put safety screens round it. Use hydrogen from a cylinder. Put a delivery tube at A. Collect the gas in test tubes. Test it until there is no explosion.

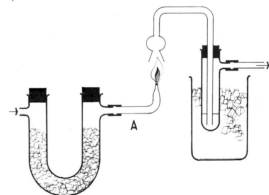

Now put a jet at A. Light the hydrogen. Gently turn on the pump. Collect liquid in the tube cooled by ice. Test to see if it is water.

1. Add the liquid to white copper sulphate.
2. If there is enough, find its boiling point.

The U-tube holds anhydrous calcium chloride. It dries the hydrogen. Why is drying needed?

> *Grace got so excited when*
> *The balloon was filled with hydrogen,*
> *She didn't let go of the string.*
> *Grace and balloon were on the wing*
> *So fast that no one could have stopped her.*
> *They brought her down by helicopter.*

The uses of hydrogen

The uses of any substance depend on its properties. Hydrogen is lighter (or less dense) than air.

Balloons filled with hydrogen will rise through the air. Weather balloons can do this. They have been used to carry instruments into the air. They measure factors such as wind speed and air pressure. These affect weather.

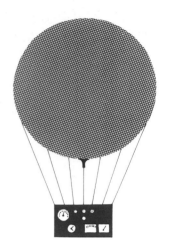

Airships are balloons fitted with engines and steering. They were first used early in this century. The later ones were rigid. That is, they had a metal frame covered with material. They were moved by propellors and driven by diesel engines.

These airships carried people and goods. The Germans called them Zeppelins. They used them in the first World War.

Britain and America also built airships. The British R34 crossed the Atlantic in 1919. It was the first to cross from Britain to USA.

In time bigger and better airships were built. Two famous British ones were the R100 and R101. The Germans built a very large one in 1936. They called it the Hindenburg.

All airships and balloons have one fault. The hydrogen which lifts them burns. If air mixes with hydrogen a light will explode it. Accidents to airships could be disastrous.

A hydrogen balloon.

A weather balloon being launched.

The burnt out remains of the R101.

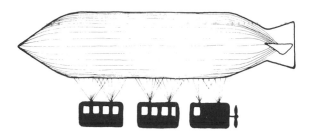

The R101 made its maiden voyage in 1930. It set out for India. It hit a hillside in France. An explosion and fire followed. The airship burnt out. Forty-eight people died.

The Hindenburg burnt out, too. It was destroyed by an explosion in 1937. It was caused by a bomb. Many lives were lost as a result.
Aircraft became bigger and faster. They could carry more fuel. So they could travel farther without stopping. Airships became less useful. They now use helium. This is a light gas, too, but it does not burn.

Barrage balloons were used in the last war. They were to protect places from air attack. The attacks were made by low-flying aircraft. They used bombs and machine guns.

The balloons were filled with hydrogen. This took them into the air. They were moored by steel cables. These made air attacks difficult. However, hydrogen burns. Many balloons were shot down in flames.

Hydrogen has other uses. Vegetable oils are easy to obtain. They come from soya beans, coconuts and ground nuts. The oils are not suitable as food.

A barrage balloon.

Hydrogen can convert oils to solid fats. Fats mixed with other food stuffs make margarine.

Ammonia is a very important substance. It is used in making fertilizer, explosives and nitric acid. Huge amounts of it are made from hydrogen and nitrogen.

$$\begin{matrix} \text{hydrogen} \\ + \\ \text{nitrogen} \end{matrix} \quad \rightarrow \quad \text{ammonia} \quad \begin{matrix} \rightarrow \text{fertilizers} \\ \rightarrow \text{nitric acid} \\ \rightarrow \text{explosives} \end{matrix}$$

Hydrogen is also used in the manufacture of methanol. This is one of the fuels used in rockets.

The first US space shuttle took off in April 1981. It had a crew of two and carried liquid hydrogen and oxygen. At take-off the hydrogen was burnt in the oxygen. In $8\frac{1}{2}$ minutes, 1500 tonnes of hydrogen was used up, or about 3 tonnes per second. It produced heat and steam. The huge thrust from this launched the shuttle.

Amazing Grace returns.

One more property of hydrogen

Calcium burns brilliantly in air. This means it must be very good at taking oxygen from the air. It can even take oxygen from water. This was how we discovered hydrogen.

Calcium + hydrogen oxide → calcium hydroxide + hydrogen

This is a chemical change. It is also a battle for oxygen. In water, hydrogen is joined to oxygen. Calcium takes away the oxygen. Calcium wins. It is more active than hydrogen.

Some games are like this. Football is a battle for the ball. George has the ball. Willie takes it away from him. Willie wins the ball. He is more active than George.

But hydrogen burns in air. It must be quite good at taking oxygen. Can it win oxygen from other metals?

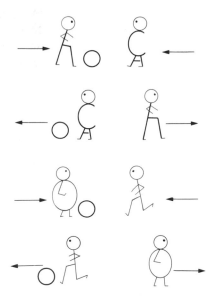

*Experiment 3.7 To test how active hydrogen is

Use a test tube with a hole at the closed end. Put a layer of black copper oxide in it. Keep it away from the stopper. Hold the tube in a clamp. Pass hydrogen through it for some time.

Light the hydrogen at the hole. Gently heat the copper oxide. Start heating at the stopper end.

copper oxide

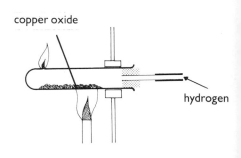

hydrogen

Is there any change in the black oxide?
Is any other substance formed in the tube?
What is left in the test tube?
Why is the hydrogen not lit at the start?

Copper oxide is copper combined with oxygen. It turns into an orange-pink substance. This might be copper. If it is, hydrogen has taken oxygen from the copper.

A mistiness appears in the tube. This could be water. If it is,
copper oxide + hydrogen → water + copper

Hydrogen has won. It has robbed copper of its oxygen. Hydrogen is more active than copper. It takes oxygen from it. Calcium is more active than hydrogen. It takes oxygen from it.

Perhaps we can work out a league table. The most active element would be at the top. The others would be placed in order of activity. Can you put calcium, copper and hydrogen in order? We can add others to the list. Where will magnesium be?

We can play a match between magnesium and hydrogen. This time give the ball (sorry, the oxygen) to magnesium. Find out if hydrogen can take this oxygen away.

Plan the experiment: What two substances must we start with?
Where shall we play the match?
Can you guess the result?

*Experiment 3.8 To find which is more active, magnesium or hydrogen

Repeat Experiment 3.7. Put magnesium oxide in the tube. Pass hydrogen through. First heat gently, then strongly.

Is there any sign of change in the tube?

No water forms in the tube. The white oxide shows no change.
No new substance is formed. No chemical change takes place.
Hydrogen does not rob magnesium of oxygen.
Magnesium wins.
Where will magnesium be in the activity list?

Summary

When fuels burn, water is formed. So water may be an oxide. Some metals burn well in oxygen. They may be able to take oxygen from water. Calcium and magnesium can. Calcium reacts faster. A gas is formed. It is called hydrogen.

calcium + water → calcium hydroxide + hydrogen

Pure hydrogen burns quietly. Water forms. Both changes show that water is hydrogen oxide. Hydrogen mixed with air can explode. This is why a lighted splint gives a 'squeaky pop'.

Calcium and magnesium are more active than hydrogen. They take oxygen from it. But hydrogen reacts with hot copper oxide. It takes oxygen from it. Hydrogen is more active than copper.

copper oxide + hydrogen → copper + water

Taking oxygen from a substance is **reduction**. Hydrogen can **reduce** copper oxide to copper.

Zinc and dilute hydrochloric acid give hydrogen. It is collected over water. Balloons filled with hydrogen rise. It is less dense than air. Hydrogen is used to make other substances.

Copy this out. Fill in the blanks.

Water is hydrogen _____. Calcium will _____ with water. The final solution turns litmus _____. It contains an _____. The word equation is

calcium + water → calcium _____ + _____

Hydrogen can be made from an acid and a _____. Dilute _____ and _____ can be used. The gas is collected in gas jars 'over _____'. It is tested with a _____ splint. It gives a small _____ or _____ pop, if mixed with _____. If it is _____ hydrogen burns quietly.

Hydrogen is _____ _____ than air. Balloons filled with it _____. Hydrogen takes _____ from hot copper oxide. The oxide is _____ to copper.

copper oxide + _____ → copper + _____

Elements can be put in order of how _____ they are. The most _____ metal comes at the _____.

Magnesium is **below calcium** but **above hydrogen**.

This is what you ought to know

In this book we have begun to study Chemistry. We have found answers to many questions.

1. *What is science?*
In science we find out about ourselves and our world. To do this we ask questions. We answer them by using observation and experiment.

2. *What is chemistry?*
Chemistry is a branch of science. The world is made of 'stuff'. This stuff is called **matter.** In chemistry we find out about the different kinds of matter.

3. *What is a substance?*
A **substance** is one particular kind of matter. Sugar is a substance. So is iron. Water is a substance too. We recognize a substance by
(a) what it looks like (its **appearance**)
(b) what it can do (its **behaviour**).
Its appearance and behaviour are its **properties.**

4. *How do we discover its properties?*
To show its true properties a substance must be **pure.** This means taking any other substances away from it. We can never get a substance 100% pure.

5. *How do we separate substances?*
A substance may dissolve in a liquid. If it does, it is **soluble.** The result is called a **solution.** The liquid is called the **solvent.** The dissolved substance is called a **solute.** It may be a solid, liquid or gas.

To get pure solvent from a solution, **distil** it. To get pure solute from a solution, **evaporate** it. The solute may appear as crystals. To get the solid crystals **filter** the liquid. A centrifuge can be used to do this. It also separates an insoluble substance.
Chromatography separates substances of the same kind.

6. *How can we test if a substance is pure?*
Simply measure its **boiling point.** Its **melting point** can also be used. An impurity raises the boiling point. It lowers the melting point.

7. *What is a chemical change?*
Substances are used to make new ones. Any change which forms a new substance is a **chemical change.** If no new substance forms the change is a **physical** one. A chemical change is also called a chemical **reaction.**

8. *What is an experiment?*
We find answers to questions. Some answers come from **observation.** This means what we see, hear and feel. For other answers we need to do something. What we do is called an **experiment.** It must be planned with care. Above all it must be safe to do.

9. *What is a theory?*
What we find out needs to be explained. We need to know 'Why?' and 'How?'. The explanation is called a **theory.** It suggests other questions to ask.

10. *What is burning?*
Heat causes physical and chemical change. In air, some substances burn with a flame. They use up part of the air. This part is **oxygen.** The burning substance and oxygen react. The new substances formed are called **oxides.** Each contains an element and oxygen. Air is roughly $\frac{1}{5}$ oxygen and $\frac{4}{5}$ nitrogen. Other gases are present in small amounts.

11. *What are the properties of oxygen?*
Oxygen is a gas. Substances burn more brightly in it than in air. It relights a glowing splint. It is made by heating some oxides. The oxide splits up or decomposes:

$$\text{mercury oxide} \rightarrow \text{mercury} + \text{oxygen.}$$

12. *What are the properties of oxides?*
A **metal oxide** may dissolve in water. If it does, the solution turns litmus **blue.** It contains an **alkali.** The family name of alkalis is **hydroxide.**

Non-metals form oxides. Most of them dissolve in water. The solutions turn litmus **red.** Each solution contains an **acid**. Its name shows which non-metal it came from. Sulphuric acid contains the non-metal sulphur.

13 .*What is neutralization?*
Acids react with **alkalis**. Correct amounts of each give a **neutral** solution. In it litmus is **mauve** in colour. It contains new substances. One is a **salt**. Universal Indicator has a range of colours. Its neutral colour is **green**.

14. *What are the properties of hydrogen?*
It is a gas. It is less dense than air. It rises through the air. Pure hydrogen burns quietly. Air and hydrogen form an explosive mixture. The substance formed is water. Water is hydrogen oxide. Hydrogen takes oxygen from some metal oxides. The metal is formed. Taking away oxygen is **reduction.**

15 *How can we sum up a chemical reaction?*
Hydrogen **reduces** copper oxide. Copper and water are formed. Tests prove that these are the products.

   copper oxide + hydrogen → copper + water

This is a **word equation.** It is a useful summary. Adding oxygen is **oxidation.** A new substance is formed. Copper oxide oxidizes hydrogen to water.

16 *Does our air remain constant?*
Fuels burn. Animals breathe. Both take oxygen out of the air. Both put carbon dioxide into the air. Plants take carbon dioxide from the air. Their roots take in water. From them they make starch and oxygen.

   carbon dioxide + water → starch + oxygen

The oxygen passes into the air. This change is called **photosynthesis.** If both changes balance, air is constant.

Questions

1. Read each sentence. Choose the *best* answer from **A, B, C, D** or **E.**
   Salt dissolves in water. We say it is:
   A insoluble;   **B** soluble;   **C** solution;   **D** solvent;   **E** dissolvable

   Water boils at: **A** 0 °C;   **B** 100 °C;   **C** 212 °C;   **D** 105 °C;   **E** 32 °C

   Substances dissolved in water make its boiling point, compared to normal:
   A higher;   **B** the same;   **C** lower

   A metal burns in air. In oxygen it would burn
   A more brightly;   **B** less brightly;   **C** not at all;   **D** feebly;   **E** slowly

2. Fill in the blank spaces.

   magnesium + oxygen → magnesium _____

   calcium + water → calcium _____ + _____

   The last solution turns litmus _____. It contains an _____. Its full name will be calcium _____.
   Sulphur is a non-metal. It burns in _____. Its oxide dissolves in _____. The solution turns litmus _____. It contains an _____. Acids react with _____.

   Name two changes which take oxygen from air.
   What are the products formed?
   Green plants take a gas out of air. Which one?
   What do they take in through their roots?
   What two substances are formed in their leaves?
   Write a word equation for this change.

3. Three jars contain different gases.
   Each is tested with a lighted splint.
   In A the splint burns brighter. The gas is _____.
   In B there is a squeaky pop. B is _____.
   In C the splint goes out. C is _____ or _____.
   Which gas, A, B or C, will you test with lime water?
   Which will you test with a glowing splint?

Metals

Hydrogen reacts with heated copper oxide. A liquid forms in the test tube. It makes white copper sulphate blue. It boils at 100 °C. These tests show that it is water.

An orange pink powder is left in the tube. It looks like copper. It has the colour of a new penny.

copper oxide + hydrogen → (copper?) + water

How can we find out if the powder is copper? What are the properties of metals?

Metals are shiny. The shine is called metallic lustre. Is the powder shiny? No, but it may be tarnished. Polish it! Have you ever tried polishing a powder? We need to know some other properties of metals.

What others do you remember? The uses of a substance depend on its properties. Make a list of the uses of copper from the pictures.

Copper is used in making coins. Observation shows that they sink in water. They are 'heavy'.

A boat with a copper hull.

Experiment 4.1 To find out about the heaviness of substances

Take cubes or lumps of substances. Use as many metals and non-metals as you can. Put each one on a balance. Read off its mass.

Measure the side of each cube in centimetres. Another way is to take a measuring cylinder. Half fill it with water. Read the level of the water. Put in the lump. Read the water level again.

The balance gives the mass of the lump. This is the amount of matter in it. It is in grams (g), or kilograms (kg).

The amount of space a lump takes up is its volume. This is measured in cubic metres (m^3) or cubic centimetres (cm^3).

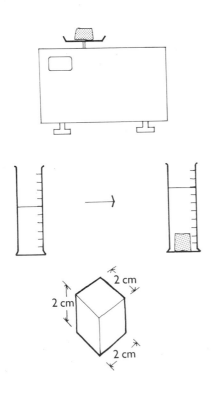

Work out the mass of 1 cm^3 of each substance. For example:
The volume of the cube shown is 8 cm^3. Its mass is 80g. 8 cm^3 have a mass of 80g. 1 cm^3 has a mass of 10g. The mass of 1 cm^3 is called the density.

A copper roof and lightning conductor.

A telephone cable containing many copper wires.

Check your list of uses of copper.

Copper is used in roofing. It is a covering for the hulls of boats. It is used in 'silver' and 'copper' coins. This is because it is not an active metal. Air and water have little effect on it.

Copper is used in wires and cables. It acts as a lightning conductor. All these carry electric current. Substances which do this are called conductors. Those which do not are insulators.

Experiment 4.2 To test electrical properties of substances

Connect the apparatus as shown. Touch the bare wires A and B together.

Take dry lumps or cubes. Use the same ones as in the last experiment. Touch one side of the lump with A. Touch the other side with B. Press the bare wires against the lump.

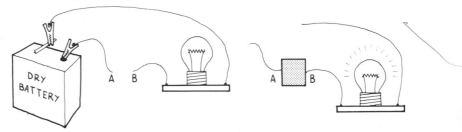

Why does the bulb light up when A and B touch?
What happens when a metal is between A and B?
Do all metals give the same result?
What happens when non-metals are used?
Do any non-metals give a different result?

When the bare wires touch, an electric current flows. It passes through the bulb. The bulb lights up.
Then a lump separated the bare ends of wire. If the bulb lights, that shows a current is passing through it. The current must be passing through the lump. The lump is a conductor. Current cannot pass through insulators. If the lump is an insulator, no current passes through the bulb. It does not light up.

Note down your results in a table:

Name of substance	metal or non-metal	element or not	mass of 1 cm³ in g	conductor or not
copper	metal	element	8.9 g	conductor

Copper is an element. It cannot be split up into other substances. Brass is a metal. However, it is a mixture. We can split it up into copper and zinc.

Metals are **conductors**. Nearly all non-metals are **insulators**. The density of a metal is usually high. Non-metals have lower densities than metals.

The Activity Series

You know some of the properties of metals now. Check what you know. Copy this out. Fill in the blank spaces.

Metals can _____ an electric current. They are _____. Density is the mass of _____. Most metals have _____ densities. They also have high _____ points. They have metallic _____.

Metals combine with oxygen to form _____. Some metals burn in air. One of them is _____. Metals which burn in air are very _____. Some cannot burn in air but can burn in _____. One is _____. It is _____ active than those which burn in air. Some metals take _____ from water. One is _____. It is very _____.

Some metals do not react with air, oxygen or water. One is _____. These metals are not _____.

All metals have similar properties. They differ in how active they are. We can put them in order of activity. The list is called the Activity Series. The most active metal is at the top.

So far we have these facts:

1. Metals heated in air (Expt 1.16) — Magnesium burns brightly. Copper turns black. Lead changes slowly to oxide. So does mercury. Platinum does not change.

2. Metals in oxygen (Expt 2.8) — Calcium burns with a blinding red flash. Magnesium burns brilliantly. Neither should be watched. Iron glows white hot.

3. Metals with water (Expt 3.3) — Calcium reacts rapidly. Magnesium reacts slowly. Look at the pictures on pages 68 and 73. Do any of these metals react with hot water?

Take group 1. Put the five metals in order. Then do the same for groups 2 and 3.

In each group the top metal is easy to place. In group 1 magnesium is most active. In groups 2 and 3 calcium is first. But in these groups magnesium is second. So the list begins:

calcium
magnesium
?
?
platinum

The least active metal is platinum. It takes bottom place.

All the other metals react in some way. It is not easy to put them in order. We need a way of comparing these middle metals.

The middle metals

To compare metals we must test them all in the same way. We match them all against the same element. The best one to use is hydrogen.

We can compare football teams in this way. We check the results of matches against the same team. The crazy teams all played Hydrogen Athletic. Look at the results. From them put the teams in order of how good (active) they are.

Lead United 1   Hydrogen Athletic 0
Magnesium United 9   Hydrogen Athletic 0
Iron County 2   Hydrogen Athletic 0
Copper City 0   Hydrogen Athletic 1
Zinc Rovers 4   Hydrogen Athletic 1
Aluminium Town 4   Hydrogen Athletic 0
Platinum Orient 0   Hydrogen Athletic 5

Comparing metals with hydrogen

Water is hydrogen oxide. Heat turns it into steam. This is a physical change. So steam is hydrogen oxide, too.

Test each metal with steam. Some may be able to take oxygen from steam. We can compare how well they do this.

Experiment 4.3 To compare metals using steam

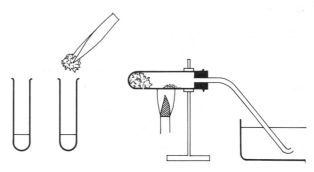

Put water into a test tube to a depth of 2 cm. Push enough rocksill in to soak up the water.

Hold the tube in a clamp. Put in a small pile of metal powder. Put it half-way between rocksill and stopper. Set up the apparatus. Collect any gas in test tubes full of water.

Heat the metal gently. If nothing happens, heat it more strongly. Test the gas for hydrogen. Examine the pile after heating. Take out the stopper as soon as you stop heating.

Your teacher may test magnesium. This helps to place others.

Which metals react with steam?
Are any more active than magnesium?

Compare the metals in two ways:
1 How fast is hydrogen formed?
2 How hot does the metal become? Does it glow or burn?
What substance is left in the tube at the end?

The bunsen flame heats the metal. It also heats the water. The water turns to steam. This passes over the heated metal. If they react, hydrogen is formed. The metal oxide is left.

Teams in order of activity
Magnesium United   **most active**
Aluminium Town
Zinc Rovers
Iron County
Lead United
Hydrogen Athletic
Copper City
Platinum Orient   **least active**

71

Putting the metals in order

Did you get the football list right? The results give the order.

Matches between metals and hydrogen give results too.

Aluminium glows brightly in steam. It takes oxygen from it. Hydrogen forms rapidly. Aluminium oxide is left in the tube.

Zinc glows red hot. Hydrogen comes off quite rapidly. Zinc oxide remains. It is yellow when hot and white when cold.

Iron glows dull red. Its oxide is blue-black. It gives a steady flow of hydrogen.

Lead has little action on steam.

These metals are more active than hydrogen. They are put above it in the list. Magnesium reacts better than any of them. So they all are below magnesium in the list.

Look at *Experiment 3.7. It shows hydrogen reducing copper oxide. Copper is less active than hydrogen. Its position is . . .?

Did you notice how fair we were? We treated all the metals alike. Each was in powder form. Each one was heated gently then strongly. So each had the same chance to win oxygen. Pass steam over zinc again. This time use *lumps* of zinc.

The list
Calcium
Magnesium
Aluminium
Zinc
Iron
Lead
Hydrogen
Copper
Platinum

Some less common metals
We shall use lithium and sodium. They are very reactive. They must not be touched. Use tongs to pick them up.

*Experiment 4.4 The action of sodium on water

Take sodium from the bottle with tongs. Blot it with filter paper. Cut off a thin slice with a knife. It should be no bigger than a match head. Drop it into water in a glass trough. Cover the trough with a safety screen.

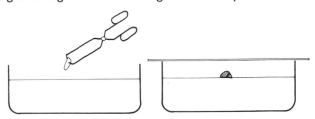

Cut a piece the size of a rice grain. All its sides must be freshly cut. Wrap it in fine wire gauze. Fill a test tube with water. Invert it in the trough. Drop the gauze into the water. Hold the test tube over it. Tap the gauze with the tube. Test the gas with a lighted splint.

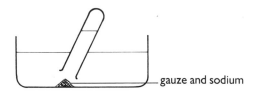

gauze and sodium

Float a filter paper on the water. Drop a piece of sodium on it. Cover the trough with a screen. Remove the screen at the end. Hold a bunsen flame in the smoke in the trough. Put a litmus paper and Universal Indicator in the water.

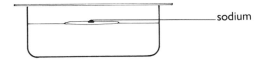

sodium

Do the same tests with the metal lithium.

Sodium has some metal properties. Which ones?
In what ways is it **not** like other metals?
Write a word equation for the reaction.
What colour is the bunsen flame in the smoke?
How does lithium compare with sodium?

Freshly cut sodium has metallic lustre. It tarnishes very quickly. This explains why it is kept under oil.
In water it melts to a silvery ball. It floats. It moves across the water. The wire gauze is used to sink it. The gas is hydrogen. The water turns litmus blue. It contains an alkali. It must be sodium hydroxide.

The reaction gives heat. This heat melts the sodium. Filter paper stops it moving. It cannot lose heat to the water. It gets very hot. It bursts into flame. It may even explode. Lithium behaves like sodium. It produces less heat. Lithium does not melt.

We cut sodium with a knife. It must be soft. It melts and floats on water. Its melting point and density must be low. Few other metals are like this.

Sodium compounds colour the flame orange-yellow. Lithium compounds colour it scarlet. Potassium compounds give a lilac flame.

Summary

All metals have similar properties. However, some are more active, or reactive, than others.

Most metals combine with oxygen. Some burn in air. Some burn only in oxygen. Some react only slowly with air. Some do not react at all.

Some metals can react with water. They take oxygen from it. Hydrogen is set free. An alkali, or hydroxide, is formed in the solution. These are very active metals such as calcium. Sodium, lithium and potassium are very active. They tarnish rapidly in air. So they are kept under oil. All these reactions produce heat.

metal + water → alkali + hydrogen + heat

Other metals react if they are heated in steam.

metal + steam → metal oxide + hydrogen + heat

The heat makes the metal burn or glow. How active they are is shown by: (a) how brightly they glow, (b) how rapidly they give hydrogen.

Finding out

1. These photographs show different metals in use. What metals are being used? What properties make these metals suitable for these uses?

2. Make a list of all the metals you can find at home. How are they protected from tarnish?

3. Find out which metals are mentioned in the Bible. What metals did the Roman Britons use? Are all these metals high or low in the Series? Find out which metals occur 'native'.

A lead roof.

A Trident aircraft made from aluminium.

The Planetarium in London has a copper roof.

Metal versus metal

In Chemistry we make substances pure. Then we find out what their properties are. By making them react we are able to make new substances.

In doing this you have learnt about many substances. You have seen many chemical changes. How can you sort it all out and remember it?

A supermarket sells thousands of things. Suppose they were dumped on the shelves anyhow. It would be hard to find a Mars bar or meat. Shopping would take hours.

Similar things are put into groups. Mars bars are with all the other sweets. Meat is with meat, not mixed with toothpaste. It is easy to find quickly the things we need.

In the same way we group substances. We divide them into classes. On page 39 we used two. A pure substance is either an element or a compound. Tin is an element, tin oxide a compound.

On page 69 we put elements into two classes. They are either metals or non-metals. We know one from the other. They have different properties.

element or compound

metal or non-metal

Putting things in different classes is called **classification.**

Metals have similar properties. However, some are more active than others. We have put them in order of how active they are. This is classification, too. We can use it to predict results. People who do football pools predict. Before the game they say what the result will be. They use the League tables.

Can aluminium take oxygen from iron? Predict! Use the Activity Series to say what may happen (p.70).

*Experiment 4.5 The action of aluminium on iron oxide.

Dry iron (III) oxide in an oven. Dry aluminium powder too. Mix equal parts of each in a crucible. Stand it on a

A supermarket.

sand tray on heat-resisting mats. Use safety screens around it.

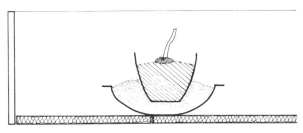

Put a small pile of sulphur on the mixture. Push a strip of magnesium ribbon into it. Light the ribbon from outside the screens.

Let the crucible cool. Scrape out its contents. Grind them to powder. Wrap a magnet in paper. Run it over the powder.

Magnesium and sulphur react. The heat this gives starts the reaction. Is it vigorous? Does it produce a large amount of heat?

Aluminium is higher in the Series than iron. Did you expect it to take oxygen from iron? That is, did you predict the result?

Taking oxygen away is called **reduction**. The magnet shows that iron is formed. We say that aluminium reduced iron oxide to iron. The reaction produces a great deal of heat. The iron formed is a white-hot liquid.

aluminium + iron(III) oxide → aluminium oxide + iron

(III) stands for three. The name is iron three oxide.

The reaction is called the Thermit reaction. It has been used in fire bombs. It can be used to weld pieces of iron. The space is filled with the mixture. Reaction is started. Molten iron is formed. This fills the gap with iron.

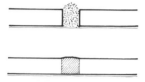

The Thermit reaction.

In the battle for oxygen, aluminium wins. It beats iron. Can magnesium reduce black copper oxide to copper? Predict!

Look at the Activity Series. Is magnesium higher than copper? Will reaction take place?

magnesium + copper oxide → magnesium oxide + copper ? ? ?

Is magnesium very much higher than copper in the Series? Will reaction be feeble, vigorous or very vigorous? Predict!

Now can you see how classification helps? Metals have the same kind of properties. They look alike. Their reactions are alike. This makes learning easier.

However, metals are not all equally active. Learn the Activity Series. It tells us how they differ. From it we can predict.

More reductions

Experiment 4.6 Can magnesium reduce copper oxide?

Dry some black copper oxide in an oven. Also dry magnesium powder. Mix small amounts of the two. Put the mixture in a crucible. Stand it on a pipeclay triangle. Put this on a tripod.

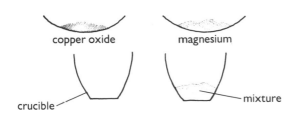

Stand the tripod on heat-resisting mats. Use safety screens. Light a bunsen burner under the crucible.

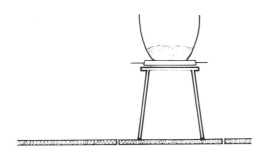

Did you predict the correct results?

Put some more mixture in a crucible. Add some magnesium oxide. Mix it in. Heat the crucible as before. Let it cool. Scrape out the contents.

Put them in a beaker. Add dilute acid. Warm the beaker. Let the magnesium oxide dissolve. Filter the mixture. Examine the solid in the filter paper.

The first reaction is almost explosive. Nothing is left in the crucible. We can only guess what is formed. Magnesium oxide added makes the reaction less vigorous. The acid dissolves it. There is a substance in the filter paper. It looks like copper.

We could predict this result. Magnesium is much higher than copper in the Series. It is much more active. It will take oxygen from copper oxide. It will do this very vigorously.

magnesium + copper oxide →
        magnesium oxide
          +
        copper

Experiment 4.7 Can iron reduce metal oxides?

Use black copper oxide, yellow lead oxide and zinc oxide. Mix each with iron powder. Fine iron filings can be used instead. Try to predict each result first. Do the lead oxide test in a fume cupboard.

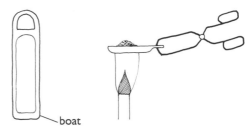

Put each mixture in a porcelain boat. Hold it in tongs. Heat it gently. If nothing happens, heat more strongly. Ceramic paper can be used instead of a boat. Wear eye shields.

Watch each mixture carefully.

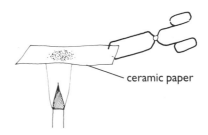

Which mixture glows when it is heated?
Which mixture shows no change?
Did you predict all three results?

The mixture of copper oxide and iron glows. Heat is being produced. This indicates a reaction. Iron is reducing copper oxide. It takes oxygen from it.

    copper oxide + iron → iron oxide + copper

Iron and lead oxide appear to react. Iron and zinc oxide show no signs of reaction.

Replacement

Iron is higher in the Series than copper. It will reduce copper oxide. Iron will reduce lead oxide too. But iron is lower in the Series than zinc. It does not reduce zinc oxide.

iron + copper oxide → iron oxide + copper

Iron has taken the place of copper. Can it do this with other copper compounds? What other copper compound do you know?

Brother George and sister Grace
Were last in a three-legged race.
Active Willie, running ace,
Pushed George out and took his place.
The chief result of this displacement
Was the startling speed at which poor Grace went.

Experiment 4.8 Replacement reactions

1. Take iron wire or a pen knife. Rub it with emery paper. Put copper sulphate solution in a beaker. Hold the clean iron in it. Leave it for a minute or two. Take it out. Look at it.

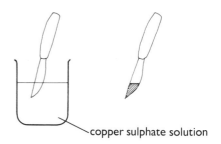

copper sulphate solution

What do you see on the iron?
What substance has formed on its surface?
Has the iron taken the place of the copper?

2. Clean a piece of copper foil. Dip it into silver nitrate solution. Leave it for a while.

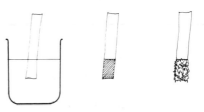

Dip clean zinc foil into lead nitrate solution. Rub lead foil with emery paper. Dip it into copper sulphate solution.

Metal forms on the copper. Is it silver?
Which is the more active metal, copper or silver?
Where should silver be placed in the Activity Series?
Does anything form on the surface of the zinc?
Which is higher in the Activity Series, lead or zinc?
Is there any sign of copper on the lead foil?

Each metal becomes coated. Copper appears on the pen knife. Iron is the more active metal. It takes the place of copper.

iron + copper sulphate → iron sulphate + copper

A white metal forms on the copper foil. It must be silver. Copper displaces silver. It is more active. Copper should be placed above silver in the Series. Lead crystals form on zinc. The surface of lead becomes copper-coloured.

Making metals

On page 73 you were asked to make two lists. One was of metals found at home. The other listed those used in early times. The Bible tells us the ones used 2000+ years ago. Metals are important.
Think of life without them. What would you miss?

Metal compounds are found in the earth. They are called ores. Many ores are oxides. Those of tin, iron and aluminium are oxides. Getting metals from ores is called **extraction.** There are many methods. The properties of the metal decide which one to use.

For example, we make iron from its oxide ore. We remove oxygen. Aluminium does this in the Thermit process. Iron mixed with aluminium oxide is formed.

iron oxide + aluminium → iron + aluminium
oxide

Getting iron from the mixture is not easy. If only aluminium oxide were a gas! It would escape from the reaction. Iron would be left on its own. Carbon burns well. This means it takes oxygen well. It may reduce oxide ores. Finish this equation.

copper oxide + carbon → copper + _____

_____

We can find out if this works. We shall use charcoal.

Experiment 4.9   Reduction with charcoal

1. Mix powdered charcoal and black copper oxide. Use one part of charcoal to two of oxide. Heat the mixture in a test tube. Light a splint. Put the flame at the mouth of the tube.

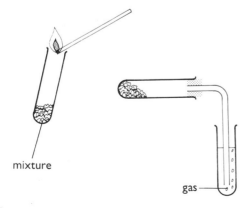

mixture

gas

Heat the mixture again. Pass any gas into lime water. Let the tube cool. Scrape out the solid.

The splint goes out. So a gas is given off.
What gases put out a lighted splint?
What does the lime water show?
The solid left in the tube is brown in colour.
Can it be copper? Write a word equation.

*2. Make a hole in a charcoal block. Put yellow lead oxide into it. Mix a little charcoal with it. Close the hole at the bottom of the bunsen.

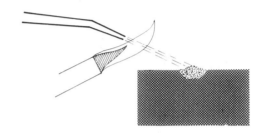

Use a mouth blow pipe. Put the jet at the back of the yellow flame. Blow the flame gently into the hole. Do this heating in a fume cupboard. Let the block cool. Remove any metal. Rub it on paper.

What does the metal look like?
Does the flame colour change?

In 1, carbon dioxide is formed. It escapes into the air. The brown substance is copper. Carbon can reduce copper oxide.
In 2, a silvery metal forms. During heating it is a liquid. On cooling it becomes a solid. It writes on paper. It is lead.

3. Try to reduce other oxides with carbon.
Use the charcoal block method. Find out if a metal is formed.
Use the test tube method. Test for carbon dioxide.

Which metal oxides are reduced?
Are the metals high or low in the Activity Series?
Did any substances on the block colour the flame?

If carbon reduces the oxide:
a) a metal forms
b) carbon dioxide is given off in the reaction.

Carbon reduces oxides of metals low in the Series.

Compounds of some metals colour the flame. We saw this in Experiment 4.4. Sodium compounds colour it orange-yellow.

We heated lead oxide on a charcoal block. It made the bunsen flame grey. Compounds of other metals give flame colours. This is useful. It shows which metal a compound contains. Try it and see!

Experiment 4.10 Flame colours

Put each substance in the edge of the flame.
Either: 1. Take a splint. Soak one end in dilute hydrochloric acid. On the wet end pick up a few crystals. Hold them in the flame.

Or: 2. Take a flame test wire. This is platinum or nichrome sealed into glass. Dip it into acid. Hold it in the flame. A clean wire will not colour the flame. Dip it in acid again. Use it to pick up some substance.

Use the substances shown in the table.

Substance	Colour of flame
sodium chloride	orange-yellow
potassium chloride	lilac
lithium chloride	scarlet
barium chloride	pale green
calcium chloride	brick red
strontium chloride	red
copper sulphate	green-blue

Summary

Metals can be placed in an Activity Series. It is also called the Reactivity Series. Metals good at taking oxygen come at the top. They take oxygen from metals lower down. Taking oxygen from substances is reduction. Adding oxygen is oxidation.

iron + copper oxide → iron oxide + copper

Heated iron and copper oxide react. The mixture glows red hot. The reaction gives heat. Copper oxide is reduced to copper. Iron has displaced copper. This puts it above copper in the Series.

Carbon is a good reducing agent. It turns into carbon dioxide. This gas escapes into the air.

tin oxide + carbon → tin + carbon dioxide

Metal compounds are found in the earth. They are called ores. Many ores are metal oxides. They can be reduced to the metal. Making metals from ores is extraction. There are several methods. The properties of the metal decide which is best. A metal displaces those below it in the Series. It can happen in solution. One example is:

zinc + silver nitrate → zinc nitrate + silver

The Activity Series can predict. It can say if a reaction is likely to happen.

Some metal compounds colour a flame. The colour is used to show which metal a compound contains.

Questions

1. An oxide is heated on a charcoal block. A silvery metal is formed. What is this chemical change called? What metal could it be? How could you test it? Write a word equation.

2. Some possible changes are listed below. Use the Activity Series to say which would happen.
   a) Copper + zinc oxide → copper oxide + zinc
   b) Lead oxide + iron → iron oxide + lead
   c) Sodium chloride + tin → tin chloride + sodium
   d) Zinc in silver nitrate solution becomes coated in crystals of silver.
   e) Aluminium reacts with copper oxide to form copper. The reaction produces a lot of heat.

3. Describe how you would prove that
   a) some blue crystals are a copper compound,
   b) magnesium oxide is not reduced by carbon,
   c) zinc comes above copper in the Activity Series.

To rust or not to rust?

Most metals tarnish in air. They lose metallic lustre. The tarnish can be rubbed off. Then the shiny metal shows again.

Some metals tarnish quickly. They may need to be kept under oil. These metals are high in the Activity Series.

Some tarnish slowly or not at all. A new penny loses its shine very slowly. Gold keeps its lustre all the time.

Iron shows its tarnish most. The tarnish is red-brown in colour. This makes it easy to see on grey iron. It is called by the special name of rust.

The layer of rust easily flakes off. This exposes the iron underneath. This rusts too. In time the iron turns into a heap of rust. This is a chemical change. What makes it happen?

From everyday life we know that rusting needs water. What other substances does it use?

Experiment 4.10 To find out if rusting uses air

Wet the inside of a test tube. Sprinkle iron filings inside it. Turn it upside down in water. Measure the height of air in the tube.

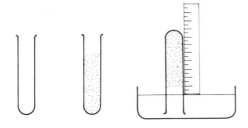

Leave it for a few days. Measure the height of air again. Test the air with a lighted splint.

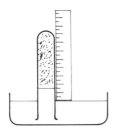

Iron rusts. Is air used by the change?
What happens to the lighted splint?
What percentage of air has gone?
What gas makes up this percentage of air?

Derelict ships rusting.

These are some results:
First height of air = 14 cm
Height after rusting = 11 cm
Height of air used = 3 cm
Volume of air is roughly proportional to height.

The lighted splint goes out. About 21% of air is used. Both results show that rusting uses oxygen. Can you guess what substance rust is?

Rusting needs water and oxygen. Rust is iron (III) oxide. It contains variable amounts of water. Iron can be protected from rusting. Water and air must be kept away from it. A protective coat is needed. This can be paint or another metal.

The manufacture of metals

The extraction of metals is a huge industry. We would miss metals. Did you work out how much? No modern transport? No cookers? No cutting tools? No wires? No electricity? No . . . . ? Can modern plastics be used instead? We shall see.

Check the Activity Series again. Silver and gold are at the bottom. This is because they are very inactive. The actual metals occur in the earth. This explains their use in early times. Man did not need to extract them. Extraction came later.

Copper and tin are low in the Series. Their ores are easily reduced. So they were discovered early in history. Bronze was also known very early. It looks like an element. It is a mixture of copper and tin.

Mixtures like this are called *alloys*. Brass is an alloy. It contains copper and zinc.

Metals high in the Series are more active. They are hard to extract. So they came into use much later. Sodium was first made in 1807. Aluminium was a rare metal until 1886. Making top metals by chemical reaction is difficult. Electrical methods are used. We shall deal with them later.

The Activity Series is a list of metals.
The properties of a metal decide its position.
Properties also decide its method of extraction.
So position is a guide to extraction method .
The Series helps us to remember and predict.

Unloading iron ore.

A bronze cannon which has been in the sea for nearly 500 years.

These coins were made 2000 years ago.

81

Breaking down other ores

Do you remember burning magnesium in oxygen? The flame is white hot. The two elements combine. They form the compound magnesium oxide. In doing so, they give a huge amount of heat (and light).

magnesium + oxygen → magnesium oxide + heat

Suppose we want the elements back. Then we must split the oxide up. This is decomposition. Magnesium and oxygen give heat as they combine. To separate them we must put that heat back.

We must heat the oxide very strongly. At 2500°C it does not decompose. It does not even melt.

Compare this with mercury. Lavoisier heated it in air for days. Mercury combines very slowly with oxygen. The compound is red mercury oxide. To get the elements back, heat it. We did this (page 38). Very gentle heat split the oxide up.

It seems that:
Some metals react vigorously giving heat. Then the compound is hard to split up.

Some metals react with difficulty. Then the compound formed is easy to decompose.

We shall try to decompose some natural ores. One of these is malachite. It is green in colour.

Experiment 4.11   To heat malachite

Take a small dry test tube. Put in the powdered ore. Heat it gently. If nothing happens, heat it strongly.

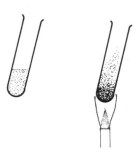

Does malachite need strong or gentle heating?
What colour is the powder left in the tube?
The powder seems to 'boil'. Can you say why?

The ore decomposes easily. It needs only gentle heat. Its metal may be low in the Activity Series. The powder 'boils'. It is blown about in the tube. A gas given off causes this. Some steam may appear.

A black powder is left in the tube. We have met three black powders. They are charcoal, manganese(IV) oxide and black copper oxide.

Willie really loves the sea,
He leaps into the waves with glee
And when the time for swimming's
* past*
It's Willie who is dragged out last.
George is the other way about,
He's last one in and first one out.
He isn't active from the start,
But Willie and sea are hard to part.

Charcoal takes litmus out of a solution (page 46). Manganese(IV) oxide is used to make oxygen (page 42). We can reduce heated copper oxide to copper (page 64). Read through these tests. Try them on the black powder from malachite. Which will you try first? Which is the powder most likely to be?

Malachite seems to give a gas. How can we find out if it does? Invent a method for doing this. Use it to collect any gas. How shall we know what gas it is?

Experiment 4.12   To find out if malachite gives off a gas

Put powdered malachite in a test tube. Fit it with a stopper. Put a delivery tube in the stopper. Heat the powder. Collect any gas in test tubes.

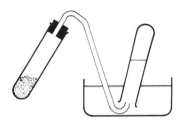

There are tests for the gases we know. They are shown below. The first test is a lighted splint.

A lighted splint in different gases	In oxygen the splint burns more brightly. Oxygen relights a glowing splint
	In hydrogen the splint goes out. The gas gives a 'squeaky pop'.
	In carbon dioxide or nitrogen the splint goes out. Carbon dioxide turns lime water 'milky'. Nitrogen does not.

Test the gas with a lighted splint. If it puts out a lighted splint, test it with lime water.

What is the effect on the splint?
Does the gas react with lime water?
What gas does malachite give?
What did the black powder prove to be?

Hydrogen is passed over the black powder. It becomes orange-pink. Water forms as well. The orange-pink substance is a conductor. It must be the metal copper.

The black powder must be copper oxide. It does not affect litmus solution. It is not a good catalyst for making oxygen.

The gas puts out the splint. It also turns lime water milky. It is carbon dioxide. We can write an equation.

malachite heated $\rightarrow$ copper oxide + carbon dioxide

Malachite is a copper ore. It is copper carbonate. Heating it gives black copper oxide. This can be reduced. It is one way of making copper. Many ores are carbonates. We shall test the ore cerrusite.

Experiment 4.13   To heat cerrusite

Heat a little powdered ore gently. If nothing happens heat more strongly. Put a lighted splint at the mouth of the tube. Squeeze the teat of a dropper. Hold it in the tube. Stop sqeezing. The dropper takes in gas. bubble this gas through lime water.

The white ore decomposes. What gas is formed? A yellow solid forms. What might it be?

Did the ore need strong or gentle heating?

It needs stronger heating than malachite does. It decomposes. It gives carbon dioxide. A yellow solid forms. It could be lead oxide. Lead is above copper in the Series. Its compounds will need stronger heating.

Experiment 4.14   To test the yellow solid

Powder the solid. Mix it with charcoal. Heat it on a charcoal block (or in a crucible). Let it cool. Take out any metal. Rub it on paper. What metal is it?

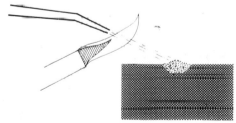

A silvery liquid metal forms. On cooling it turns solid. It writes on paper. It must be lead. Cerrusite is lead carbonate. Charcoal reduces lead oxide.

lead carbonate $\rightarrow$ lead oxide + carbon dioxide
lead oxide + carbon $\rightarrow$ lead + carbon dioxide

Lead is a soft, dense metal. It is not reactive. It corrodes only slowly. It has a low melting point. Find out the everyday uses of lead.

Chalk and limestone

Chalk is a soft natural rock. Millions of years ago tiny creatures lived in the sea. Their skeletons fell to the sea bed. Pressure turned them into chalk. Larger pressure gave limestone. Heat and pressure gave marble. They are common rocks, harder than chalk.

Experiment 4.15   To heat chalk

Put powdered chalk into a test tube. Heat it very strongly. Pass any gas into lime water.

Can you see any change in the chalk?
With very strong heating does it give a gas?
Will a direct flame heat it more strongly?

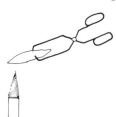

Hold a chalk chip in tongs. Put it in the hottest bunsen flame. Heat it all over for ten minutes. Stand it on wire gauze to cool. Wait until it is quite cold. Put a lump of chalk beside it on the gauze. Fill a dropper with cold water. Add drops to each cold lump. Press a litmus paper on each wet solid.

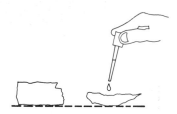

Does direct heating appear to change chalk?
Does the flame alter during heating?
Does cold water react with chalk?
Does water react with the lump left after heating?
What does the litmus paper show?

Chalk in the test tube shows no change. A few bubbles of carbon dioxide may form. The tube may melt first! Chalk is more strongly heated in the flame. The lump glows white hot. The flame may become coloured.

Water dropped on chalk soaks in. Nothing else happens. Water on the cold lump turns to steam. Heat is needed to do this. It cannot come from the lump or the water. Both are cold. It must come from a chemical change.

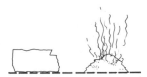

The lump crumbles to a powder. The powder turns litmus blue. It is an alkali. Which alkali is it?

Shake the crumbled powder with water. Filter. Divide the solution into two. Blow through a glass tube into one half. Add Universal Indicator to the other half.

Read page 79 again. What happens to burning calcium?

What do these tests show?

The indicator becomes violet. This shows a strong alkali. Breath contains carbon dioxide. It turns the clear liquid milky. This liquid could be lime water.

Experiment 4.16   A coloured flame from chalk

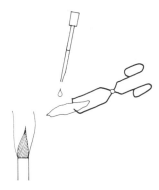

Hold a lump of chalk in tongs. Put it very close to a bunsen flame. Fill a dropper with dilute hydrochloric acid. Put drops on the chalk near the flame.

Do chalk and acid react? What is formed?
What change is there in the flame?

Chalk and acid give a gas. The flame becomes brick red. Calcium compounds give this colour. The alkali powder is calcium hydroxide. The clear liquid we made is lime water.

The limes

Experiment 4.17 Limestone and marble

Repeat Experiment 4.15. Use limestone and marble instead of chalk. Compare results for the three.

> Do marble and limestone behave like chalk?
> Can they be the same chemical compound?

Marble and limestone react exactly like chalk. they are forms of the same compound. Heating decomposes it. Carbon dioxide is formed. the compound is a carbonate, like malachite.

This carbonate is hard to decompose. Its metal must be high in the Activity Series. Its flame colour is brick red. It must be calcium carbonate.

calcium carbonate $\rightarrow$ calcium oxide + carbon dioxide

Calcium oxide is called **quick**lime. It reacts with water vigorously, giving heat. This heat turns some water into steam. The oxide crumbles to a powder. This is an alkali, calcium hydroxide.

calcium oxide + water $\rightarrow$ calcium hydroxide

Quicklime glows when strongly heated. It gives white light. This was once used for stage spotlights. Important actors were 'in the limelight'.

Calcium hydroxide is called **slaked** lime. It is slightly soluble in water, forming lime water.

Limestone is an important mineral. So is chalk. In Britain we use about 50 million tonnes a year. About one fifth goes to make quicklime and slaked lime. They are all used in the building industry.

Limestone is used in making iron and steel. It is mixed with iron ore and coke. The mixture is fed into a blast furnace. Molten iron can be run out.

Quicklime and slaked lime are used in farming. They neutralize acid soils. They break down heavy soils. This makes the soils easier to work. It also improves their drainage.

Summary

Some metal ores are carbonates. Heat decomposes them. Carbon dioxide is given off. The metal oxide remains. Some carbonates are easy to split up. Their metals are low in the Activity Series. Malachite is one of these. It is copper carbonate. Heating it gives copper oxide. It can be reduced to the metal. Metals are made from carbonate ores.

Chalk, marble and limestone are calcium carbonate. It is hard to decompose. The oxide is quicklime. Quicklime and water react. Calcium hydroxide forms. It is called slaked lime. Heat is also formed. All these substances are important. They have many uses in building, farming and industry.

Questions

1. I have a piece of white rock. I heat it very strongly. It gives carbon dioxide. A white lump is left behind. The white lump reacts with water. It crumbles to a powder. This turns litmus blue. Steam is also formed. The rock gives a flame colour. It is pale green. Explain all these changes. What substance is the rock?

2. Write down answers which fill the blank spaces. Chalk is the compound calcium _____. It needs very _____ heat to decompose it. _____ oxide is left. The gas _____ _____ is given off. Calcium oxide is called _____lime. Calcium oxide and water react. The new substance turns litmus _____. It is an _____ called calcium _____. Its everyday name is _____ _____.

Some people enjoy being in the limelight.

85

Chemistry and building

Have you ever seen a house being built?
First of all the shape of the house is marked out on the ground. Then trenches are dug. This gives a base for every wall.
Each trench is filled with concrete. These are the foundations. On them the walls are built.

The walls may be of local stone. Bricks, breeze blocks and concrete blocks are more common. Each block is bonded to the ones round it. Thin layers of mortar are used to do this.

On top of the walls a wooden frame is put. Tiles or slates are put on it to form the roof.

Spaces are left in the walls. These are for doors or windows. Glass is used in window spaces. The walls are rough on the inside. They are covered with a smooth layer of plaster.

Clay, limestone and sand are used in building. Limestone is, like chalk, calcium carbonate. Sand is silicon dioxide. It belongs to the same family as carbon dioxide. Carbon dioxide gives carbonates. Sand forms silicates.

Clay is a mixture of aluminium silicate and sand. Mixed with water, it gives a paste like mud. **Bricks** and **tiles** are shaped from this. They are then baked in an oven called a kiln.

Cement is made from finely powdered limestone and clay. This mixture is heated in a rotating furnace at about 600 °C. It is cooled. It is ground to a fine grey powder. This is cement.

Cement is mixed into a paste with water. It is moulded into shape. **Concrete** is made of cement, small stones and sand. It is mixed with water and shaped like cement. Concrete and cement both dry out. They set to a hard mass like rock.
Concrete can be set round a steel frame. This makes it much stronger. It is called reinforced concrete.

Foundations and the start of walls.

These buildings in Stamford are made of local stone.

Buildings can be made in many different styles.

Glass is made in huge flat sheets.

Glass blower

Mortar is a mixture. It has six parts sand, one part slaked lime and one part cement. It gives a paste with water. This sets hard like cement. But it is much cheaper. Slaked lime is an alkali. It combines with carbon dioxide from air. This slowly forms limestone, making the mortar stronger.

Glass is a mixture. It is made from sand, soda and limestone. These are measured out and mixed. The mixture is heated in a furnace to about 1500 °C. It melts. The liquid is glass. It is as runny as treacle. Lumps of it, called gobs, are shaped in moulds. Air blows the glass into shape.

Flat glass is made by two methods. Liquid glass flows over rollers. These are cooled by water. Molten glass is also floated on liquid tin. This gives it a very smooth surface. Other substances are put in glass mixtures. They alter its properties to match the uses.

The mineral gypsum is calcium sulphate. It loses water when heated, like copper sulphate. The substance formed is **Plaster** of Paris. It is made into a smooth paste with water. It is spread on walls and ceilings. It slowly dries. It combines with the water. This gives gypsum again as a smooth plaster.

Putty is used to fit glass into wooden frames. It is chalk mixed with linseed oil. The linseed oil reacts with air. This makes the putty harden out.

Carbon dioxide

In the last experiments we heated three carbonate ores. We shall now find the effect of adding acids to them.

Experiment 5.1 The reaction of acid with carbonate

Malachite is copper carbonate.
1. Put a small amount of it in each of three test tubes. Add dilute hydrochloric acid to one. Add dilute nitric acid to the next. Put dilute sulphuric acid in the third. Find out what gas is formed.

The drawings show ways of passing the gas into lime water.

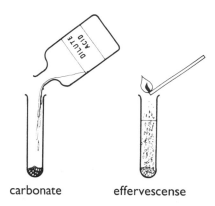

carbonate   effervescense

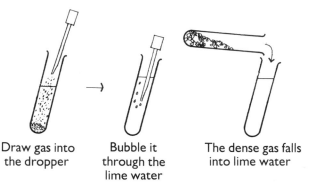

Draw gas into the dropper | Bubble it through the lime water | The dense gas falls into lime water

What tells you that a gas is formed?
Did you test it with a lighted splint?
Did you pass the gas through lime water?
What colour is the solution in each case?

2. Add the same dilute acids to (a) cerrusite, (b) chalk, (c) marble. Test the gas.

Do they all produce carbon dioxide?
Which of the carbonates give gas rapidly?
Which carbonate and acid give a steady stream?

The mixtures froth and bubble. This effect is called **effervescence**. It shows that a gas is being formed. We used the usual tests on it.

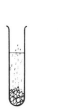
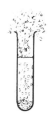

A lighted splint in different gases	In oxygen the splint burns more brightly. Oxygen relights a glowing splint.
	In hydrogen the splint goes out with a squeaky pop.
	In carbon dioxide or nitrogen the splint goes out. But carbon dioxide makes lime water milky.

All carbonates react with acid to give carbon dioxide. Some reactions soon stop. The powders give gas at a very rapid rate. They may foam out of the test tube. Lumps such as marble give gas more slowly.

We shall find out more about carbon dioxide. To do this we need gas jars full of it. For this we need a steady stream of the gas. Lumps of marble provide a steady flow.

Marble and sulphuric acid soon stop reacting. So we shall use dilute hydrochloric acid.

The properties of carbon dioxide

Experiment 5.2 To make and test carbon dioxide

To collect the gas in gas jars

To collect it in test tubes

Hold a lighted splint at the top of the gas jar. When the jar is full of gas the flame goes out.

1. Shake carbon dioxide with water. Add a litmus paper. Put in some Universal Indicator.

What do the indicators prove?

2. Invert a test tube of the gas in (a) water, (b) dilute alkali solution.

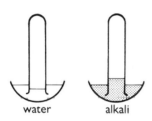

water   alkali

In which does the gas dissolve more?

3. Light a candle. Hold a gas jar of carbon dioxide over it. Take the cover off the jar.

 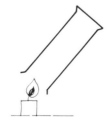

What does this test tell us about the gas?

4. Light a piece of magnesium ribbon. Hold it well down in a gas jar of carbon dioxide.

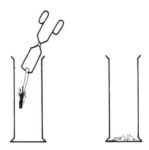

Does magnesium go on burning?
Is the powder left in the jar one substance?
What could this substance be?

5. Bubble carbon dioxide through lime water (calcium hydroxide solution) for some time.

Carbon dioxide is more dense than air. It falls into a gas jar. It pushes out the air. It fills the jar.

It also falls over the candle. It puts the candle flame out. A candle cannot burn in carbon dioxide.

Magnesium burns in carbon dioxide. A white powder is left. It must be magnesium oxide. It has black specks in it. These could be particles of carbon.

$$\text{magnesium} + \frac{\text{carbon}}{\text{dioxide}} \rightarrow \frac{\text{magnesium}}{\text{oxide}} + \text{carbon}$$

In water, carbon dioxide forms a weak acid. It is called **carbonic acid**. This is why it dissolves well in alkalis. It reacts with them. Its salts are carbonates.

carbon dioxide + water → carbonic acid

The lime water test

Were you surprised when the lime water 'turned milky'? Surely not! We have used this test many times.

Carbon dioxide in lime water forms white chalk. The chalk floats in water. It makes it 'milky'.

$$\text{calcium hydroxide solution} + \text{carbon dioxide} \rightarrow \text{chalk} + \text{water}$$

Did you expect it to become clear again? Putting in more carbon dioxide made it clear. If it is clear, the chalk has gone. But chalk does not dissolve in water. It must have formed a new substance.

$$\text{chalk} + \text{water} + \text{more carbon dioxide} \rightarrow \text{new substance}$$

The new substance must be soluble in water.

Experiment 5.3 To heat the clear solution

Half fill a small beaker with lime water. Bubble carbon dioxide through it. Stop when the liquid becomes 'milky'.

clear → milky

milky → clear again

Bubble more carbon dioxide through it. Wait until it becomes clear. Gently heat this until the water boils. Watch what happens.

Now it's cloudy...

...now it's not!

Boil it for four or five minutes. Let it cool. Pour away the water. Leave the solid in the beaker. Add dilute acid to it.

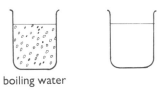

boiling water

Do you live in a chalk or limestone area? If so, boil half a beaker full of tap water.

> What happens to the clear solution on boiling?
> Did you see bubbles of gas form in it?
> Does the white solid react with the acid?
> Does tap water behave in the same way?

When the clear solution is heated, bubbles form in it. This gas must be carbon dioxide. A white solid forms. This dissolves in acid. It also gives gas. It must be chalk.

Tap water from a limestone area gives the same results.

The clear solution contains a new substance. It is called calcium hydrogencarbonate. The equation is:

$$\text{chalk} + \text{water} + \text{carbon dioxide} \rightarrow \text{calcium hydrogencarbonate}$$

The 'milky' liquid contains chalk. Chalk is calcium carbonate. It reacts with carbon dioxide. The new substance is called calcium hydrogencarbonate. It dissolves. This explains why the solution becomes clear.

Heat reverses the change. Carbon dioxide bubbles appear. Chalk forms again. The solution is 'milky'.

The uses of carbon dioxide

In water carbon dioxide forms a weak acid.

About 1 cm³ of gas dissolves in 1 cm³ of water. Much more gas will dissolve under pressure. The solution is pleasant to the taste. Fizzy drinks contain carbon dioxide under pressure. The gas escapes when the bottle is opened.

Carbon dioxide is also the gas in wines like champagne. It is formed when wines and beer are made. Some of it remains in the liquid.

Carbon dioxide puts out burning substances.

Two effects put out a fire. One is cooling it. The other is keeping air from it. Water does both. However, it leaves an awful mess at the end.

A liquid carbon dioxide cylinder does both. It gives cold gas. This cools the fire and keeps out air. Unlike water, it leaves no mess. However, it is a gas. This means it does not stay over the fire. It is most useful for small fires.

The gas is a non-conductor. It can be used safely on electrical fires. It blankets the fire better if it is mixed with foam. Powder fire extinguishers contain sodium hydrogencarbonate. It is heated by the fire. It gives carbon dioxide.

Solid carbon dioxide is very cold. It turns to gas at − 78 °C. It is used in refrigeration. It is colder than ice. It leaves nothing behind

after use. This why it is called 'dry ice'. This is important in food transport. It is used when low temperatures are needed.

Fizzy drinks contain carbon dioxide.

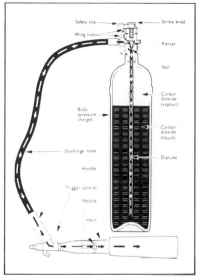
A fire extinguisher.

Questions

1. Fill in the blanks.

   Carbon dioxide is made by adding _____ acid to _____. It is _____ dense than air. This means it _____ into a gas jar. It puts out a lighted _____ or burning _____. It is used in fire _____. It could not put out a fire of burning _____.

   Magnesium burns in carbon dioxide. The equation is:
   magnesium + carbon dioxide → _____ + _____. Carbon dioxide _____ in water. The solution is a weak _____. It is used in making _____ drinks. Solid carbon dioxide is very _____. It turns to gas at − _____ °C. It is used to transport _____.

2. A test tube contains a gas. It is hydrogen, oxygen, carbon dioxide or nitrogen. How would you find out which? Describe what you might see.

Water for us

When we need water we turn on a tap. Clean water comes out. It has no harmful substances in it. It has no bacteria which carry disease.

Where does our water supply come from? It first falls as rain or snow. This soaks into the earth. Some of it may collect underground. Much of it runs into streams. These grow into rivers. Rivers flow into the sea.

Sea and rivers evaporate. Water vapour rises into the air. As it cools it turns back into drops of water. We see these drops as clouds. If the drops get big enough they fall as rain. If it is cold enough, ice crystals fall as snow. We are back to where we started. This series of changes is called the **Water Cycle**.

We need water all the time. We store it in reservoirs. Even these may run dry in hot weather. They are filled by rivers and rain. The water from them is made clean. It passes into pipes called mains. These take it to the taps.

All the used water goes down the drain. It is cleaned up and put back into the rivers or the sea.

Underground water is also used. A hole is drilled down to it. A pipe is passed down the hole. The water can then be pumped up the pipe.

We can add these three changes to the Water Cycle. We shall split the Cycle into sections. Then we can study each section in more detail.

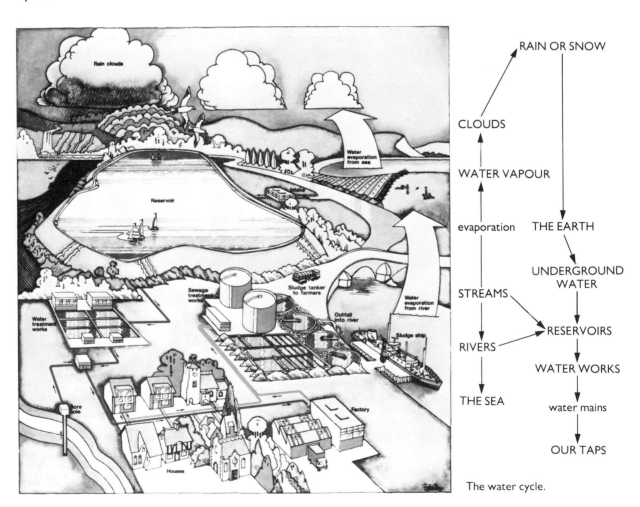

The water cycle.

From rain to reservoir

Water is a solvent. Many solids, liquids and gases dissolve in it. As rain falls it will dissolve gases out of the air. One of these gases is carbon dioxide.

The rain soaks into the earth. It runs over rocks. Two common ones are chalk and limestone. The map on page 100 shows where they occur in Britain. They are both calcium carbonate.

On page 89 we passed carbon dioxide into milky lime water. Three substances met in the beaker. They were chalk, water and carbon dioxide.

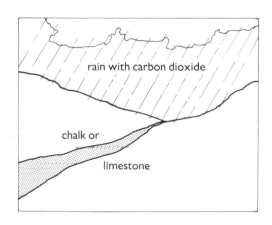

The chalk dissolved. It did this by turning into a new substance. This new substance is soluble. It is calcium hydrogencarbonate.

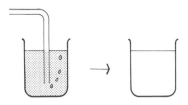

Rain falls on chalk or limestone. The same three substances meet. Some chalk or limestone dissolves. This puts calcium hydrogencarbonate into rivers and streams.

The rain and rivers feed reservoirs. These supply our tap water. In many parts of Britain tap water contains calcium hydrogencarbonate.

$$\underset{\textbf{(rock)}}{\text{calcium carbonate}} + \underset{\textbf{(rain and rivers)}}{\text{water} + \text{carbon dioxide}} \rightarrow \underset{\textbf{(tap water)}}{\text{calcium hydrogen-carbonate}}$$

Rain and rivers pass over other rocks. Some of these dissolve. Calcium sulphate is one of them. Tap water will have this in it too.

We drink tap water. We wash in it. We boil it. It evaporates if we leave it in the air.
What effect will the calcium compounds have?
How will calcium compounds affect these changes?

Drinking

Our bodies need calcium compounds. Having them in our drinking water is useful. They are used in forming bones and teeth. They may also give water a more pleasant taste.

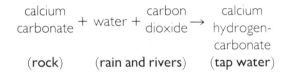

AVERAGE ANALYSIS
DISSOLVED SOLIDS PARTS PER MILLION

Calcium Carbonate	100
Magnesium Sulphate	38
Magnesium Chloride	32
Sodium Chloride	26
Silica	10
Sodium Nitrate	6
Potassium Nitrate	2
Organic Carbon	0.1
Fluoride	0.1
Iron	0.05
Trace Heavy Metals	less than 0.01
pH 8.0	

The label from a bottle of Malvern Spring Water.
The water contains more calcium carbonate than any other substance. It is present as calcium hydrogen carbonate.

What about washing?

We wash with soap and water. We use detergents to wash clothes and dishes.

Experiment 6.1   The effect of soap on water

We shall compare distilled water and tap water. We need to measure how much soap we use. The best way to do this is to use soap solution. This is soap dissolved in ethanol and water.

Use a small conical flask or a large test tube. Put in 10 cm³ of distilled water. With a dropper add one drop of soap solution. Stopper the tube or flask. Shake it.

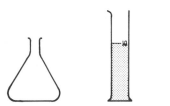

The bubbly froth which forms is called lather. If it lasts for two minutes, add no more soap. If not, add another drop of soap solution. Count the drops needed to get a lasting lather.

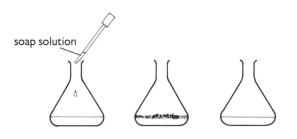

soap solution

Do this again with 10 cm³ of your tap water. Do it with 10 cm³ of tap water from a limestone area. If you cannot get this, make hard water by the method shown.

Which needs least soap to form a lasting lather?
Which water stays clear and which becomes cloudy?
What happens to tap water when soap is added?
Does water from a limestone area give the same result?

To form a lather:
distilled water needs little soap and the water remains clear.
tap water from a limestone area needs much more soap. It becomes cloudy. A white curd or scum forms in it.

Hard water

Soap in water forms a lather. The bubbles show that there is spare soap there. In some water it is hard to get a lather. It needs a lot of soap. We say that this water is **hard.**

Hard water contains dissolved substances. These react with soap. They produce a white scum. It goes on until the dissolved substances are used up. More soap can then form a lather.

Some water needs little soap. A lather forms at once. No substances react with the soap. No scum forms. The water stays clear. This water is **soft.**

To make hard water:
Shake calcium sulphate in water. Filter it. Pass carbon dioxide into lime water until it is clear. Mix equal volumes of the two solutions. This mixture behaves like water from a limestone area.

Willie said to his Dad, 'Would you rather
Shave in hard or soft water, Father?'
Dad said, 'Don't be dumb!
Hard water forms scum
And needs far more soap for a lather.'

The two kinds of hardness

Hard water wastes soap. It uses it to form scum. Scum is a nuisance. It sticks to baths and wash basins.

Distilled water would be best to wash in. It has no dissolved substances to react with soap. No soap is wasted in making scum. Lather forms straight away.

To make distilled water needs heat. This makes it expensive. Can you think of a cheap substitute? Water evaporates. The vapour rises into the air. Cooling turns it back to water. This falls as rain. Catch it before it has a chance to dissolve rock. **Rainwater is soft** (and cheap, too).

Carbon dioxide makes lime water cloudy then clear. In the solution is calcium hydrogencarbonate. Can you remember what happened when we boiled it? Hard tap water contains calcium hydrogencarbonate, too.

Experiment 6.2 The effect of boiling on hardness

Put hard tap water into a beaker. Gently boil it for ten minutes. Let it cool. Put 50 cm³ of it in a conical flask.

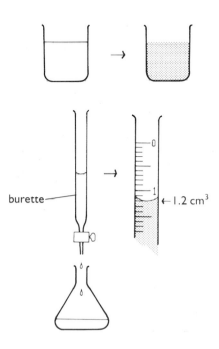

Lather produced by soap in hard and soft water.

Add soap solution from a burette. Stopper the flask. Shake it. Add just enough soap to get a lasting lather. The burette measures how much soap this needs.

Do the same with 50 cm³ of hard tap water. Test 50 cm³ of distilled water in the same way.

Water from a limestone area gave these results:

	tap water	boiled tap water	distilled water
Soap used	15.0 cm³	5.4 cm³	1.4 cm³
Water was	very cloudy	fairly cloudy	clear

Read the results. Use them to fill the gaps.

In distilled water soap forms no _____. All the soap is used in forming _____. Tap water needs about _____ times as much soap. Most of this soap is wasted in forming _____. Boiled water needs less soap than _____ water. This means boiling must have removed some _____.

Boiled tap water needs more soap than _____ water. This means that it is still _____. Boiling gets rid of calcium hydrogen _____. It turns this substance into _____. Tap water must have a second substance dissolved in it.

How do soap and detergents work?

Experiment 6.3 To compare soap and detergents in hard water

Put hard water into two test tubes.
To one add three drops of washing-up liquid.
To the other add 1 cm³ of soap solution.
Stopper both tubes. Shake them.

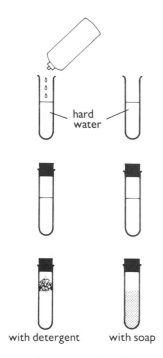

with detergent   with soap

In what ways are the two results different?

Soap is made by heating oils or fats with alkali. The alkali used is sodium hydroxide. Glycerol is formed as well.

oil or fat + sodium hydroxide → soap + glycerol

A simple name for soap is sodium stearate. In soft water it froths, or makes a lather. In hard water it forms scum first.

calcium compounds (hard water) + sodium stearate (soap) → calcium stearate (scum) + sodium compounds

Enough soap turns all the calcium compounds into scum. Adding more soap will now make a lather. A lather shows that soap is there, ready to wash with.

The word detergent comes from a Latin word. It means 'to clean'. Soap and detergents do the same job. They both remove dirt.

Detergents are made from oils too. They are heated with sulphuric acid. Detergents are not sodium compounds. They do not react with calcium compounds. So they do not form scum in hard water. They can remove dirt straight away.

Later on we shall make soap and a detergent.

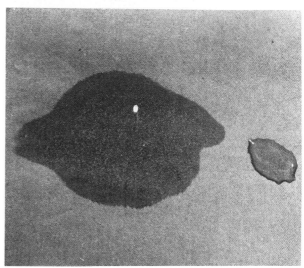

Drops of water on a cloth. The one on the left has detergent on it.

Experiment 6.4 To find out how a soap and detergent work

1. Fill a dropper with distilled water. Put a drop on the back of your hand. Tilt your hand. Carefully put a drop on a piece of cloth.

Half fill small test tubes with distilled water. Put drops of detergent in one. Put soap solution in the other. Put drops of these on the back of your hand. Put drops of each on cloth.

Do the drops of water wet your hand evenly?
Do they spread out and wet the cloth?
Do soap and detergent make a difference?

96

2. Set up three test tubes as shown. In one put two drops of liquid detergent. In the second put soap solution. Leave the third.

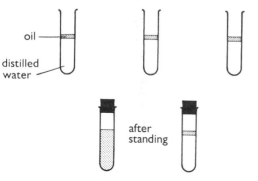

Stopper all three tubes. Shake them well. Leave them to stand. Pour the liquids away.

In which tubes do oil and water mix?
In which do they stay mixed after standing?
When each liquid is poured off is any oil left?

3. In the beakers are strips of cloth. Each is smeared with oily dirt. Rub them hard in the liquids. Pour the liquids out of the beakers.

Dispersing an oil slick at sea, using detergent from the barrels.

In which beakers is dirt removed?
In which do oil, dirt and water mix?
Is the oily dirt poured away with the liquid?

Our skins produce an oily substance. Oil and water do not mix. This is why water does not 'wet' our hands. It does not spread evenly over the skin. The oily layer prevents this.

The oily substance on the skin collects dirt. Oily dirt collects on clothes too. Water alone will not remove it. The oily layer and the water do not mix.

Soap and detergents alter this. They help oil and water mix. Experiment 6.4 part 2 shows that they stay mixed. Pouring the liquid away leaves no oil behind. After washing we rinse. This takes oil, dirt and soap (or detergent) away.

Summary

Oil and water form separate layers. There is a boundary between them. Shaking them breaks the boundary. On standing, they separate into two layers again. The two liquids do not mix.

Most dirt is oily. Water does not mix with it. So washing in water only will not work. Most of the dirt remains on skin or clothes. It may even rub off on the towel!

Soap and detergents make oil and water mix. They break the boundary between them. Oil, dirt and water form one liquid. Rinsing removes it. The dirt goes too. Clothes or skin are left clean.

Hard water has dissolved substances. They react with soap until they are used up. Scum is formed. Then more soap makes oil and water mix. Cleaning takes place. But soap has been wasted. Detergents do not react with hard water. They mix oil and water straight away. No detergent is wasted. No scum forms. Lather shows that soap or detergent is there in the water.

Question

Most dirt is oily. Explain why. How does soap help in washing? Hard water interferes in two ways. Explain them. How do soap and detergent differ?

Getting rid of hardness

Calcium compounds make water hard. So do magnesium compounds. To make water hard they must be dissolved in it.

There are two kinds of hardness. One is caused by calcium hydrogencarbonate. Boiling the water turns this into chalk. Chalk does not dissolve in water. So the hardness has gone.

Hardness removed by boiling is called **temporary hardness**.

Boiling does not remove the second kind. It is caused by calcium sulphate and magnesium sulphate. These dissolve in water. They did so when the water ran over rocks.

Hardness not removed by boiling is called **permanent hardness**.

We can get rid of both kinds. We use substances called water softeners. Washing soda, ammonia and permutit are softeners.

Experiment 6.5 To test water softeners

Take seven conical flasks. Put 50 cm³ of hard tap water in each. Add drops of washing soda solution to six of them. Give each flask a different number of drops: 1, 2, 4, 6, 8 or 12 drops. Let the flasks stand overnight.

Do this again. Add ammonia solution instead of soda. Use 2, 4, 6, 8, 12 or 16 drops.

Let hard tap water run through permutit. Collect it. Put 50 cm³ into a conical flask.

Run soap solution into each flask. Find out how much is needed to get a lasting lather.

Tap water from a limestone area gave the results below. They are also shown as a graph.

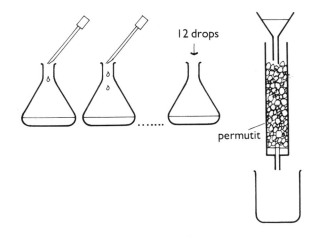

12 drops

permutit

Using soda solution (sodium carbonate):

Drops added	none	1	2	4	6	8	12
Soap needed in cm³	15.0	13.1	10.3	4.1	3.1	2.7	1.7

Distilled water needed 1.3 cm³.
Boiled tap water needed 5.4 cm³.
Water passed through permutit needed 1.7 cm³ of soap solution.

Using ammonia solution:

Drops added	none	2	4	6	8	12	16
Soap needed in cm³	15.0	13.9	9.1	6.7	6.3	5.6	5.5

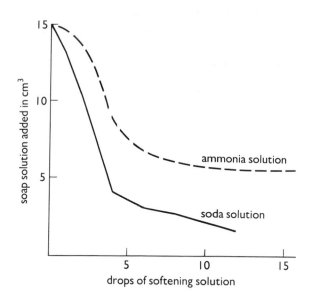

soap solution added in cm³ (vertical axis)
drops of softening solution (horizontal axis)

ammonia solution

soda solution

Look at the results and the graph. Answer these questions.

How much soap does distilled water need?
Which other solutions need almost the same amount?
What softeners get rid of all hardness?
How much soap does boiled tap water need?
Which other solution needs almost the same amount?
Which hardness does this softener remove?
Why is permutit the best softener?

Distilled water uses little soap. Two samples of water need almost the same amount. One had run through permutit. Washing soda had been put in the other. 12 drops of solution were needed.

Natural permutit is a sodium compound. The water contains calcium compounds. As it runs through, an exchange happens. Calcium and sodium 'swop'. The water loses its calcium compounds. Permutit takes out all hardness. It changes into a calcium compound. The process is called 'ion exchange'.

Washing soda is sodium carbonate. It reacts with calcium compounds. Chalk is formed in the water.

$$\underset{\text{(soda)}}{\text{sodium carbonate}} + \underset{\text{(hard water)}}{\text{calcium compounds}} \rightarrow$$

$$\underset{\text{(chalk)}}{\text{calcium carbonate}} + \underset{\text{(soft water)}}{\text{sodium compounds}}$$

The water is less hard. The graph shows it needs less soap. Enough soda takes out all hardness.

Boiled water needs less soap. Water with ammonia needs the same amount. Both methods get rid of temporary hardness only.

Questions

1. Using soap, rainwater is better for washing than tap water. Give two reasons for this.

2. Water from three towns is tested with soap. The amounts needed, in cm³, are shown:

Place	Tap water	Boiled tap water
Bultown	10	4
Cowtown	4	1
Dogtown	6	6

What kinds of hardness are present in each?

3. Copy this into your book. Fill in the blanks.

Calcium and _____ compounds make water hard.
To do so they must be _____ in the water.
Removing these compounds makes water _____.
Boiling the water removes _____ hardness.
This kind of hardness is caused by calcium _____.
Boiling turns it into calcium _____ (chalk). Chalk does _____ make water hard. It does not _____.
Boiling water does not remove _____ hardness.
It is caused by calcium and magnesium _____.
The two types of hardness are _____ and _____.
Adding ammonia gets rid of _____ hardness only.
All hardness is removed by _____ and _____.
The best softener is _____.

Things to do

1. Collect rainwater. Test its hardness with soap. (Try soap solution at school, flakes at home.)

2. Find out if bath salts soften water.

3. Try washing a greasy plate with cold water. Then use hot water. Finally use water and soap or detergent.

4. Test liquid shampoo.
   Dilute some of it with ethanol. Fill a dropper with this liquid. Measure 100 cm³ of hard water. Put it in a conical flask. Add a drop from the dropper. Shake the mixture. Find how many drops are needed to form a lather.
   Repeat the test with 100 cm³ of a) tap water, b) distilled water, c) rain water.
   What does the liquid shampoo contain?

99

Evaporating hard water

In limestone areas tap water contains calcium hydrogencarbonate. Boiling the water turns this into chalk or limestone. The same change happens when the water evaporates.

The map shows the areas in England and Wales where water is hard. Do you live in one of them?

Limestone has dissolved for millions of years. This has formed deep valleys and caves. The valleys are called gorges (see page 18). The passages down to the caves are called potholes.

Water drips from the roofs of the caves. Each drop evaporates a little. It leaves a small deposit of limestone. This slowly grows. After millions of years a pillar of it has formed.

not, they need 'de-scaling' regularly. Water softeners are also used in the home.

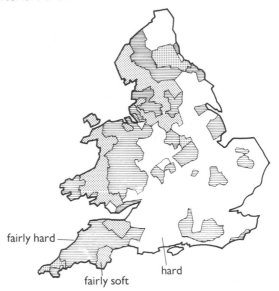

fairly hard

fairly soft

hard

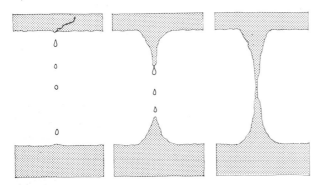

The drops fall to the floor of the cave. As each evaporates, limestone forms. This grows, too. A pillar grows upwards. The two pillars may meet. Water runs down. The whole thing grows thicker.

The pillar on the ceiling is called a stalactite. The one on the ground is a stalagmite. They are often coloured. The colour comes from other substances in the water.

Dripping taps often form limestone in the same way. You may see it on baths and wash basins.

Heating hard water gives the same effect. Limestone collects in kettles. It forms a solid layer of 'fur' or 'scale'. It also forms in hot water pipes. They may be blocked by it.

This is one more disadvantage of hard water. Water is made soft for use in boilers in industry. If

Stalactites and stalagmites in an underground cave.

Scale in a water pipe.

Reversible reactions

Water, chalk and carbon dioxide react. Calcium hydrogencarbonate is formed in the water.

$$\text{calcium carbonate} + \text{water} + \text{carbon dioxide} \rightarrow \text{calcium hydrogen-carbonate}$$

We heated this solution. It gives back calcium carbonate, water and carbon dioxide.

$$\text{calcium hydrogen-carbonate} \rightarrow \text{calcium carbonate} + \text{water} + \text{carbon dioxide}$$

This change can happen either way. Substances react to form new ones. The new ones can react. They give the substances we began with.

This kind of change is a **reversible reaction**.

Permutit is an ion-exchange compound. It contains sodium compounds. Substances like it can be man-made. They are 'ion-exchange resins'. When hard water passes through them, we get:

$$\underset{\text{(hard water)}}{\text{calcium compounds}} + \underset{\text{(exchange resin)}}{\text{sodium X}} \rightarrow \text{sodium compounds} + \text{calcium X}$$

The sodium compounds go on with the water. All the hardness has been taken out. But the resin slowly becomes a calcium compound. It can no longer soften water.

A strong solution of salt is passed through it.

$$\underset{\text{(salt)}}{\text{sodium chloride}} + \underset{\text{(used-up resin)}}{\text{calcium X}} \rightarrow \text{calcium chloride} + \underset{\text{(resin)}}{\text{sodium X}}$$

In modern softeners this happens automatically. The resin is renewed. The change is reversible. Calcium chloride is washed out by the water.

We have used an arrow to show chemical change. To show a reversible change we use half arrows.

For example: mercury heated in air gives mercury oxide. Heating the red oxide gives mercury and oxygen.

$$\text{mercury} + \text{oxygen} \rightleftharpoons \text{red mercury oxide}$$

Scale in a hot water tank.

An automatic water softener which uses Permutit.

Clean and pure

Our water comes from rivers and streams. It is stored in reservoirs. Some of these are natural. Some are man-made. A dam built across a valley makes a reservoir. So does a natural lake. Some man-made ones are huge concrete tanks.

We also get water from underground. In some places it rises as natural springs. In others we need to bore down to it. Pipes are put into the boreholes. Water can be pumped up and stored.

All water needs some treatment. How much treatment depends on where it comes from. Water from underground springs is clean. River water can be very dirty. It all goes through a water works. Here it is made pure. It is then piped to our taps.

It passes through wire screens. These take out floating objects. The water may contain muddy solids. It is left to stand in tanks. The solids settle out.

All water drains through filter beds. These are large tanks. At the bottom are stones and gravel. Over them is a layer of fine sand. The bed acts as a filter. It takes out mud and similar matter. The beds are regularly cleaned and re-made.

The water is now clear and clean. It may still contain bacteria. Some bacteria cause disease. As the water leaves the works, chlorine is added. It kills bacteria. The amount used is small. It is not enough to harm us.

We expect water when we turn the tap. So it must be stored above the area it serves. In flat areas water towers do this. Pumping is needed to fill them.

A natural reservoir.

Building a man-made reservoir. Note the man on the roof.

Cleaning filter beds.

Chlorinators add chlorine to water to kill germs.

Where does the waste go?

A family uses about 250 litres of water a day. We drink it. We wash in it. We use it to flush WCs. The used water is called waste. It goes down the drain. It takes with it all we put in it. This includes soap, detergent, dirt and organic matter.

Factories also use water. Their waste contains other substances. Some of them may be harmful.

All this waste is called **sewage.** It was once put straight into the nearest river. Near the coast it went into the sea. It made the rivers and sea dirty and smelly. It helped to spread disease.

Bacteria in the water broke down the sewage. It used up oxygen in the water. Animals and plants living in it died. They had no oxygen.

Now sewage passes through a sewage works. It is screened to take out large objects. Grit and sand are removed. They can be used for hole-filling.

The sewage then stands in large tanks. Solids settle to the bottom. This is crude sludge. It is taken out and treated separately.

The liquid sewage goes to aeration tanks. Bacteria grow in these tanks. They feed on the waste matter and destroy it. As in the river, this process needs oxygen. So air is bubbled through. This is called aeration.

The sewage has gone. The water contains only bacteria. These settle out in more tanks. They are put back into the aeration tanks. They are used again. The water is clean enough to be put into the river or the sea.

The crude sludge is also treated with bacteria. No air is allowed in. Breakdown of the sludge takes about 30 days. It gives off gas. This is used as a gas fuel in the works. The liquid is dried and used to improve the soil.

The London river is called 'Father' Thames. In the nineteenth century it was very dirty. Sewage from all over London poured into it. The cartoon shows a famous scientist. He was Michael Faraday. He wrote to the Times about the filthy river.

Beddington sewage works from the air.

FARADAY GIVING HIS CARD TO FATHER THAMES;
And we hope the Dirty Fellow will consult the learned Professor.

103

Pollution

What is it?

We make and use harmful substances. Some of them get into the air. Some find their way into rivers. The rivers take them down to the sea. Our air and water are not clean. This is pollution.

Air pollution

This is caused mainly by burning fuels. The products of burning pass into the air. They may include dust, smoke and grit. We can see these easily. Harmful gases may be mixed with them. We can't see them. We breathe them in without knowing it.

Water pollution

Sewage and other waste may get into water (page 102). Many substances are used in farming. Rain may dissolve them. They pass into rivers. Factories produce waste. It may reach a river by accident. Animals and plants live in water. They may be harmed by this pollution.

Is it a new problem?

No! It is as old as the burning of fuel. A law was passed in 1273 AD. It banned the burning of coal in London. In 1307, people still broke this law. They were punished with 'great fines'. In 1661 there was still smoke. It made London like a 'suburb of Hell'. It was blamed for all kinds of disease such as 'coughs and cathars'.

How big is the problem?

Solid fuels burning use oxygen from air. Carbon dioxide and water form. They are both harmless. They pass into the air. They do not pollute it.

Suppose the fuel has too little air. Some carbon monoxide forms instead. Solid fuels often contain sulphur. This burns to form sulphur dioxide. Carbon monoxide and sulphur dioxide are gases. We cannot see them. Both of them are poisonous.

With even less oxygen, carbon is formed. We see it as smoke. Some parts of the fuel cannot burn. They pass into the air as dust or grit.

Transport uses mainly liquid fuel. It burns. The main products are water and carbon dioxide. Some carbon monoxide is mixed with them. Most petrol contains lead compounds. Burning it puts lead compounds into the air. They are poisonous too. Nitrogen compounds burn. They give nitrogen oxides.

The effects of pollution

Smoke and solids in air absorb light. They make places darker. Windows are dirtier. We get less sunlight. We use more electricity.

We breathe air. Our lungs take in the solids and harmful gases. More people will suffer from lung disease. People are more often away from work.

Before the Clean Air Act (1910).

After the Clean Air Act (1970).

Smog and fog are more common. Fog is a mixture. It is tiny drops of liquid. They contain solid particles. Smog is a mixture of smoke and fog. It holds the solids and harmful gases. They increase until the smog goes. Both are bad for our lungs.

In 1952 London had 4 days of dense smog. Traffic was halted. Accident figures were higher. There were 4000 more deaths than usual.

Grit and dust fall out of the air. It falls most on cities. Each square km gets about 200 tonnes a year. It makes buildings, clothes and people dirty. We use more soap and detergent.

Acid rain

Rain falls through the air. Some gases dissolve in it. Sulphur dioxide and nitrogen dioxide do. In water they form acids. The result is acid rain. It attacks buildings. It dissolves the stone. It makes metals corrode. It harms plant and animal life.

The cost of pollution

Pollution damages people and buildings. More cleaning and repairs are needed. It causes fog and smog. More people are ill and more working time is lost. It may cost us as much as £500 million a year. Plant life suffers too. Plants prefer clean air.

The Clean Air Act of 1956 banned black smoke. Solids had to be taken out of it. Fuel had to be used more efficiently. The figures show the smoke put into air in a year. They are in millions of tonnes. They show the change the Act made. In 1938, 2.7; in 1966, 1.3; in 1972, 0.5.

Foggy day in London—Tower Bridge.

Ideas in chemistry

How well have you spent your time in Chemistry?
What have you learnt so far?

First We have looked at substances. We have done experiments.
Experiment and observation have given us facts. Here are some
of them. They were chosen at random.

Ice melts at 0 °C. Water boils at 100 °C. Oxygen relights a glowing splint. Salt has cubic crystals. Plants use carbon dioxide and produce oxygen. A light will make a hydrogen-air mixture explode. Water is hydrogen oxide. Mercury oxide breaks down into mercury and oxygen. Calcium burns in air. A white powder is formed.	Facts

Second Hundreds of facts like these have led to ideas.
Chemistry deals with substances and how they change.

Boiling turns water into steam. Steam is the same substance as water. In physical changes no new substance is formed. So boiling is a **physical change.** It is also a **change of state.**	Physical change
Calcium burns in air. It combines with oxygen. The product is calcium oxide, a new substance. So this is a **chemical change** or **reaction.** Calcium oxide is a **compound.** It contains calcium and oxygen. A chemical change forms at least one new substance.	Chemical change
Heat decomposes mercury oxide. It splits up into simpler substances, mercury and oxygen. They do not **decompose.** They are elements. An **element** cannot be split up into simpler substances. The world is made up of **92 elements.**	Elements
All substances have **properties.** These are their appearance and behaviour (what they look like and what they do).	Properties
Elements fall into two main classes. Those in one class are shiny. They conduct a current. Most have high melting points and densities. These elements are metals.	Metals
Elements in the other group are **non-metals.** They differ from metals. They have no metallic lustre. Most are non-conductors. Their melting points and densities are usually low.	Non-metals

These are properties of the elements. No new substances are being formed. So they are **physical properties.** Elements and compounds react. Properties are what they do. Their reactions form new substances. they are called **chemical properties.**

Physical properties
Chemical properties

Elements are placed in two classes. This is an example of **classification**. We can classify inside a group. All metals are similar. However, some react more strongly than others. We can put them in order of activity. Classification helps us to learn and remember.

Classification

Activity Series

Some substances speed up chemical changes. They are called **catalysts**. They are still there at the end of the reaction.

Catalysts

Questions

Look at the photographs. Find words to fill the blank spaces below. Write them down in your book.

1. The balloons contain gas. The gas could be _____ or _____. A flame under the balloon makes it explode. The gas must be _____. The explosion is a _____ change. The equation for it is:
   _____ + oxygen → _____.
   The balloon rises because _____ is _____ dense than air. This is a _____ property.

2. This is an _____ platform. It drills for a liquid called _____ _____. The liquid is piped ashore. It goes to an oil _____. It is a _____ of many liquids. They are separated by _____. Two liquids formed are _____ and _____. Both burn. They are liquid _____. They burn to form _____ and _____ dioxide. This change takes _____ out of the air. Green plants put _____ into the air. They take out _____ _____. This is _____synthesis.

3. Metal compounds in the earth are called _____. Malachite is one. It is _____ in colour. On heating it turns _____. It gives off a _____. This turns lime water _____. It is _____ _____. The black powder may be copper _____. Hydrogen is passed over it while it is heated. The solid formed will _____ an electric current. The mass of 1 cm^3 of it is 9 grams. It is likely to be a _____. It has a _____ density.

   copper _____ + hydrogen → _____ + _____

   Removing oxygen from a substance is _____. Hydrogen is a _____ agent. It reduces copper o_____ to form the metal, _____. Charcoal is a form of the element _____. It reduces lead oxide to _____. This is a metal. It has a _____ density. It has metallic _____.

Movement of matter

*Experiment 7.1 Movement in liquids and gases

1. Set up two gas jars as shown. Slide the lids out from between them. Count up to five. Separate the jars. Test each for carbon dioxide.

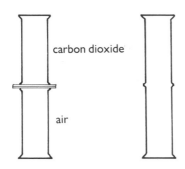

carbon dioxide

air

Which test did you use for carbon dioxide?
In which gas jars is it found?
Can you explain this result?

Take fresh jars of each gas. Do the same experiment again. This time, put the jar of air on top.

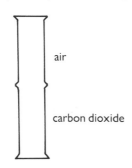

air

carbon dioxide

Is there carbon dioxide in the top jar?
If so, how does it get there?

2. Grease round the edge of a gas jar lid. Put three drops of bromine on the middle of it. Cover it at once with a gas jar. Put a white screen behind it. Let it stand. Record results.

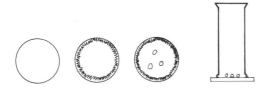

What happens to the drops of bromine?
What other change happens in the jar?
Can you explain why this happens?

In 1. there is carbon dioxide in both jars. We know carbon dioxide is more dense than air. We expect it to fall out of the top jar. However, it also rose from the bottom jar into the top. It did so in five seconds. How do we explain this?

In 2. the liquid bromine evaporates. Its vapour is orange-brown. A vapour is gas coming from a liquid. It slowly moves up the jar. In the end it fills it.

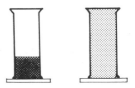

Do gases and vapours stretch like elastic?
Or do they grow upwards like plants?
Or are they made up of tiny moving bits?

Movement like this happens every day. We can smell good food from a long way away. The perfume of flowers comes in through the window. Why and how does it happen?

We can smell good food from a long way away!

Did you guess the answer? All substances consist of small bits, or particles. They are called atoms and molecules. Those in gases move freely. They travel fast in all directions.

*Experiment 7.2 To find out if particles all move at the same speed

Use a glass tube about 1 metre long and 3 cm in diameter. Support it in a stand.

Take a pad of cotton wool. Wet one side with concentrated ammonia solution. Wet a second pad with concentrated hydrochloric acid. Hold the pads about 3 cm apart.

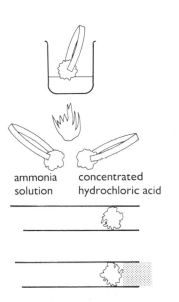

ammonia solution

concentrated hydrochloric acid

Push one pad, wet side first, into one end of the tube. At the same time push the second pad into the other end. Stopper the ends. Time how long it takes to get a result.

What happens when the pads are close?
What do you see inside the glass tube?
Did you smell any gas coming from the pads?

Each liquid gives off a gas. The gases form white smoke when they meet. A gas comes from each pad of cotton wool. The gases pass along the tube. Where they meet a smoke ring forms.

Is the smoke ring half-way along the tube?
If not, is it nearer to ammonia or acid?
Which of the two gases travels the faster?

The smoke ring is nearer the acid pad. The ammonia has travelled farther. It must have been moving faster. The gases take a long time to meet. They seem to travel slowly. Remember that the tube is full of air. It has molecules too. They are also moving.

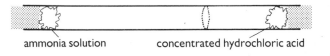

ammonia solution   concentrated hydrochloric acid

Willie, a rapid runner, reckons
To run this street in 20 seconds.
Brother George is not so swift. He
Runs the street in over 50.
Start one at each end of the street.
Who's run the furthest when they meet?

Atoms and molecules

In Chemistry we ask questions about substances. Many of them begin with 'What . . . .?'
   What causes hardness of water?
   What can we use to soften it?
To answer each question we plan experiments. The results give facts. The facts give the answer.

We all ask 'What . . . .?' questions. They are part of everyday life. A learner motor cyclist may ask:
   What does the cheapest petrol cost?
   What happens if I press the starter?

Some learners go on to ask 'How . . .?' and 'Why . . .?'.
   How does petrol reach the engine?
   Why is it hard to start on cold days?

In Chemistry we ask 'How . . .?' and 'Why . . .?' questions. The answer is called a **theory**. It explains the facts we know. It suggests new questions to ask.

We ask 'How does matter move of its own accord?' Science chose the 'moving bits' theory. It says **'All matter consists of tiny moving particles.'**

Take a strip of copper. Cut it in half. Cut one of the halves in two. This gives a quarter. Cut it in two. Let the cutting go on. Can it go on for ever?

'Why doesn't the engine work?'

If this table is made up of tiny atoms...

... why doesn't it collapse in a heap?

NO! We come to a 'smallest bit' of copper. This 'smallest bit' is called an **atom** of copper. The idea that atoms exist is very old. Democritus suggested it over 2000 years ago. The name atom comes from Greek words. It means 'not cuttable'. Ideas about atoms make up the **Atomic Theory.**

But no one had seen an atom. No one could say what held atoms together. The theory fell flat. In 1803, John Dalton revived and added to it. We now know that all matter is made up of atoms.

An atom is the smallest piece of an element. The smallest piece of carbon is a carbon atom. The smallest piece of oxygen is an oxygen atom. Carbon burns in oxygen. Carbon and oxygen atoms meet. They join together. Bigger particles form. Each of them is a molecule of carbon dioxide.

A molecule is two or more atoms bonded together.

The theory explains

We know that matter moves. How does it do this? The Atomic Theory should explain how. It says that Matter consists of moving atoms and molecules.

In Experiment 5.2, carbon dioxide fell out of the top jar. We expected this. It is more dense than air. But it also rose out of the bottom jar. How?

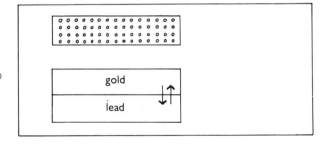

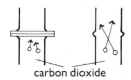

carbon dioxide

It is a gas. Its molecules move. They travel freely in all directions. Some will hit the gas jar lid. We took the lid away. Molecules which would have hit it can now go on. They pass into the top jar. We added lime water. This proved they were there.

Drops of bromine spread through the gas jar. How?

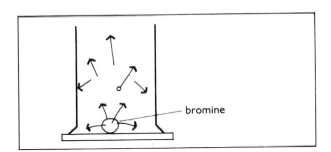

bromine

Bromine is a liquid. The molecules inside the drops move freely. They hit the surface. Some are fast enough to go through. They pass into the air as gas or vapour. They still move freely. They spread to fill the gas jar. Why does this take so long?

The theory explains other changes. A gold block is stood on a lead block. It stays there for years. The gold next to the lead is tested. It contains traces of lead. The top surface of the lead contains a little gold. Each metal has moved into the other.

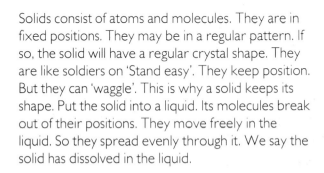

Solids consist of atoms and molecules. They are in fixed positions. They may be in a regular pattern. If so, the solid will have a regular crystal shape. They are like soldiers on 'Stand easy'. They keep position. But they can 'waggle'. This is why a solid keeps its shape. Put the solid into a liquid. Its molecules break out of their positions. They move freely in the liquid. So they spread evenly through it. We say the solid has dissolved in the liquid.

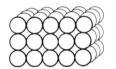

All substances spread by atoms or molecules moving. This movement is called **diffusion.** Flowers give off substances. Foods do too. The molecules of these substances diffuse. They reach our noses. They may affect nerve endings in the nose. We 'smell' them.

Atoms and molecules move. Do they all move at the same speed? Why does the movement seem so slow? How big are they? How many in a drop of water?

A theory explains. It also suggests new questions.

Atoms – are they really there?

Atoms are far too small to be seen. So are molecules. Why do we believe that they exist?

First:

We can see their effects. What they do shows. A jet plane flies very high. It is a long way off. This may make it too small to be seen. But we can hear it. We can often see its vapour trail. These effects show that it is there. We cannot see a mole under a lawn. It makes small mounds of earth. Mole hills tell us it is there. What it does shows.

Second:

The Atomic Theory can answer 'How' and 'Why' questions. It explains our results in Science. It is the only explanation of some results. One example of these is Brownian motion.

Vapour trails tell us that a plane is high in the sky.

Experiment 7.3 The Brownian motion

Set up a small glass cell. Focus a beam of light through it. Puff some smoke into the cell. Look at it through a microscope. Describe what you see.

What is it which is actually moving?
What makes this movement happen?

Smoke consists of tiny bits of solid. The beam of light acts as a searchlight. The solid bits of smoke are lit up. They look like tiny points of light. We can see them through the microscope.

They move continually. The effect was first seen in 1827. Robert Brown saw pollen grains floating on water. He looked at them through a microscope. They moved in all directions. They never stopped. Brown gave his name to this: 'Brownian motion'.

A pollen grain cannot move on its own. Neither can a solid particle in smoke. Something must make them move. Something must make them change direction too. Both changes never stop.
They have air or water round them. Both consist of molecules. These are in constant motion. Can you see how they might cause Brownian motion?

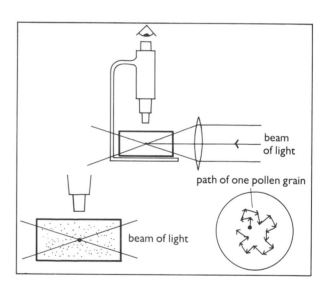

Questions

Use what you know about molecules to explain why
a) ice crystals have a definite shape,
b) water takes the shape of the vessel it is in,
c) cooling turns a vapour into a liquid,
d) cooling turns a liquid into a solid.

The picture shows a balloon. Imagine millions of people round it. They throw tennis balls at it. Suppose that more balls hit it on the left side. Which way will it move?

A second later more may hit it on the right. What happens to its movement? Will it change?

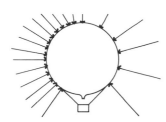

A smoke particle is like a balloon. Molecules of air are like the tennis balls. They hit the particle on all sides. By chance more hit it on one side. It moves away from that side. A moment later more hit it on another side. This keeps it moving continuously. Only the direction alters.

One more idea is important. Heat gives energy. With more energy, molecules will move faster. Only moving molecules explain Brownian motion. They explain many other results. Here are some.

Melting. We know that:
Solid ice melts at 0°C. Its molecules are in fixed postions. They move only very slightly.

Heat gives them energy. They move faster. This shakes them out of those positions. Now they can move freely. Ice has melted. 0°C is hot enough to make this happen. 0°C is the melting point.

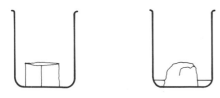

Evaporation and boiling. We know that:
Liquids evaporate. Heat makes it happen faster. Water boils at 100°C. When it boils it bubbles.

Molecules inside water move freely. Some hit the inside surface. Fast moving ones go through. They pass into the air. The water gets less. Heat makes molecules faster. More can break out into the air. The water evaporates much faster. At 100°C huge numbers escape. We see them rise as bubbles. They show that the water is boiling.

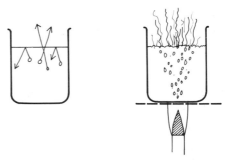

Molecules - Are they really there?

From Experiment 7.2 we know that:
Different molecules travel at different speeds. We compared these speeds. Both pads give a gas. Molecules travel fast. Why do they take so long to meet? The tube is full of air. Its molecules move too. Could you run fast through a crowd?

Willie the rapid runner reckoned
To beat his record by a second.
Alas, he had to keep on stopping.
Saturday morning crowds were shopping.
Not even Zola Budd or Cram
Could run fast through a shopping jam.

113

Atoms and molecules – how big are they?

Remember: a molecule consists of atoms bonded together.

Experiment 7.4 The mass of a molecule

Fill a small beaker with water. Take a crystal of potassium manganate (VII). Measure its mass. Put it in the water. Let it stand for some time. Stir it.

What do you see after letting it stand?
What do you see after stirring?

Fill a dropper with solution. Run drops into a measuring cylinder. Count the number needed to make 1 cm³. Look at a drop through a microscope.

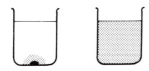

Is there potassium manganate (VII) in each drop?
Can you work out how much?

Purple solution forms round the crystal. Stirring spreads it evenly. The whole solution becomes pink. Each drop has some manganate (VII) in it.

Here are some results:
Mass of crystal   = 0.005 g
Volume of water   = 100 cm³
Number of drops in 1 cm³ = 25

So 100 cm³ of solution contains 0.005 g of solute
    1 cm³ contains 0.005 ÷ 100 = 0.000 05 g
    1 drop contains 0.000 05 ÷ 25 = 0.000 002 g
One hundredth of a drop is pink. It contains about 0.000 000 02 g. This must be one molecule at least. A molecule has a mass less than this.

Molecules and atoms must be very, very, very small.

A molecule has a very tiny mass. Suppose it is a sphere, like a tennis ball. Can we measure its diameter or thickness? Think of a rope loop on a pond. Fill a bucket with tennis balls. Pour them into the loop. Do they fill it?

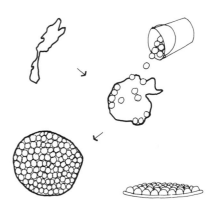

If not, push the side of the loop. It will be dented. Fill the loop with balls. It may take more than one bucket full. The layer is one tennis ball thick. The loop will not now dent.

We can fill a loop with oil molecules. The oil is stearic acid. It is in solution. The solvent is petroleum ether. 1000 cm³ contains 0.1 cm³ of oil.

114

Experiment 7.5 The diameter of a molecule

Take three watch glasses. On 1. put a drop of the stearic acid. On 2. put a drop of petroleum ether. On 3. put a drop of the solution.

Fill a large funnel with water. Tie 30 cm of cotton into a loop. Grease it slightly. Drop it on the water. Put a drop of oil solution into the loop. Tap it with a pencil. Add more drops, one at a time. Count the drops. Stop when the loop is just full.

  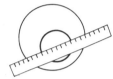

Measure across the loop in centimetres. Find out how many drops make 1 cm³ of solution. Work out the area of the circle. It is a layer of oil one molecule thick. Work out its thickness.

Which liquids evaporate away on the watch glass?
Which part of the solution evaporates on water?
Which part is left floating on the water?

Petroleum ether evaporates. The oil does not. Oil molecules move. They spread over the water. They fill the cotton loop. It makes a circle. We can work out its area. We can work out the volume of this oil.
The volume of oil = its area × its thickness

My results

Distance across cotton loop = 10 cm

Area of cotton loop
$$= 3 \times 5^2 \text{ cm}^2 = 75 \text{ cm}^2$$

1000 cm³ of solution contains 0.1 cm³ of oil

1 cm³ of solution contains 0.0001 cm³ of oil

There were 20 drops in 1 cm³ of oil solution

1 drop contains 0.0001 ÷ 20 cm³ of oil

3 drops were needed to fill the loop with oil.
They contain 3 × 0.000 005 cm³ = 0.000 015 cm³.

This is the volume of oil in the loop. Its area is 75 cm². Volume is area × thickness.

The thickness of the loop = diameter of one molecule = 0.000 000 2 cm.

The result is about one ten millionth of a centimetre.
This is one thousand millionth of a metre.

Atoms and molecules are very, very, very small.

115

More about atoms - atomic theory

Science does not stand still. New facts are always being discovered. So ideas and theories change. The Atomic Theory is a good example.

Democritus first suggested it. There were no facts to support it. No one could explain what held atoms together. So the theory failed.

It was revived in the 17th century. Gases had been investigated. Gas laws were discovered. Moving molecules could explain these laws.

In 1803, Dalton used these ideas. He added new ones. He used atoms to explain chemical change. It became 'Dalton's Atomic Theory'. Much more has been discovered since then. Compare Dalton's ideas with modern ones. You will learn more about these later.

Dalton's Atomic Theory

Every element is made up of atoms. They are small and cannot be split up.

Atoms of the same element are alike. They have the same size and mass.

Atoms of one element are different from the atoms of others. They differ in size and mass.

Modern Ideas

Every element consists of atoms. Atoms are made of tinier particles.

Atoms of the same element are alike. They may not all have the same mass.

Atoms of different elements are not alike. They all contain the smaller particles but not in the same numbers.

Atoms join together. The result is a molecule. So the smallest bit of compond is a molecule. It has atoms of different elements.

Symbols

It is difficult to show an atom on paper. Dalton used a different sign for each kind. In 1803 not many elements were known. Only a few signs had to be learnt. Now we know over 100.

Simple molecules were easy to show. A carbon dioxide molecule has one carbon atom. It has two oxygen atoms. The drawing shows this.

Large molecules were more difficult. A molecule of sugar has 12 carbon atoms. It has 22 hydrogen and 11 oxygen atoms. The drawing shows this.

But this method is clumsy. A much simpler one was found. The first letter of the name of the element is used. O stands for an oxygen atom. A hydrogen atom is H. Each letter is called the symbol of the element. It stands for one atom. Choose symbols for nitrogen and carbon.

a carbon atom ●
a hydrogen atom ☉
an oxygen atom ○

a carbon dioxide molecule

The symbol for carbon is C. Then what do we use for calcium? For cobalt? For copper? Their names begin with C. We add a second letter of the name.

The symbol for calcium is Ca. For Cobalt it is Co. In Dalton's day people still used Latin. The Latin word for copper is cuprum. Its symbol is Cu.

The masses of atoms are tiny. Dalton was sure no one would ever measure them. He decided to compare the masses of atoms. He used compounds to do it.

Take black copper oxide. Its smallest part is a molecule. It contains copper and oxygen atoms. Dalton guessed it had one of each. In symbols it would be Cu joined to O. Its **formula** is CuO.

A formula shows what atoms a molecule contains.

In Experiment 4.6 we reduced copper oxide. We can use it to compare the masses of atoms.

Experiment 7.6 To compare the masses of copper and oxygen atoms

Dry some black copper oxide in an oven. Gently warm a reduction tube. Let it cool. Measure its mass in grams. Put into it a layer of the oxide. Measure its mass again. Hold it in a clamp and stand.

Pass hydrogen through it for some time. Light the hydrogen at the hole. Gently heat the oxide. Start at the stopper end. When it has all changed let it cool. Keep hydrogen passing through while it cools. Find the mass of the tube and its contents.

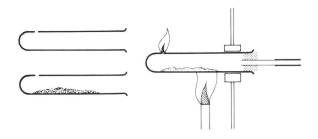

Hydrogen reduces copper oxide. It takes oxygen away from it. Copper is left. Water is also formed.
copper oxide + hydrogen → copper + water

This is a set of results.
Mass of the tube and copper oxide = 18.97 g
Mass of the tube and copper = 18.17 g
Mass of the empty tube = 14.97 g

Answer these questions. Use your own results. If you have none, use the results above.

What is the mass of copper oxide?
What is the mass of oxygen taken from it?
How much copper remains in the tube?

Removing oxygen makes the mass less. It falls from 18.97 g to 18.17 g. The mass of oxygen is 0.80 g.

The copper formed is $18.17 - 14.97$ g $= 3.20$ g

In copper oxide there are 3.20 grams of copper
0.80 grams of oxygen

With 0.8 g of oxygen there are 3.20 g of copper

With 1 g of oxygen there would be $\dfrac{3.20}{0.80}$ g of copper

1 g of oxygen combines with 4 g of copper

How many atoms of copper are there in 4 grams? We do not know. Call this number of atoms x.

Dalton guessed : 1 atom of copper to 1 atom of oxygen.

Cu Cu Cu Cu Cu Cu . . . . 4 grams of copper
O O O O O O . . . . 1 gram of oxygen

Cu atoms must have four times the mass of O atoms.

Dalton compared other atomic masses.
The hydrogen atom has the smallest mass.
The mass of the oxygen atom is 16 times as big.

The formula for copper oxide is CuO. It contains one atom of copper and one of oxygen. A molecule of carbon dioxide contains one atom of carbon. It has two oxygen atoms. Its formula is written CO_2. The number after a symbol shows how many atoms. So the formula of sugar is $C_{12}H_{22}O_{11}$. What atoms are there in the formula H_2SO_4?

Relative atomic mass and formula

Look at the photograph.
The mass of the marble is 5 g. Its symbol is M.
The mass of the golf ball, symbol Gb, is 45 g.
This is 9 times as big. If M is 1, Gb will be 9.
The cricket ball mass is 155 g. Call its symbol Cb.
Its mass is 31 times the mass of the marble.

If we make M = 1, then Gb = 9 and Cb = 31.

These numbers **compare** the three masses. They are 'relative masses'. We can do the same with atoms.

The hydrogen atom has the smallest mass. Call it 1.
The mass of an oxygen atom is 16 times as big.

If H = 1, then O = 16.
A copper atom has 4 times the mass of an oxygen atom.

If H = 1 and O = 16, then Cu = 64.
These numbers are the **relative atomic masses.** Others are shown in the table under RAM. They are given to the nearest whole number. We shall use them to find the formula of a compound.

Element	Symbol	RAM
Aluminium	Al	27
Argon	Ar	40
Bromine	Br	80
Calcium	Ca	40
Carbon	C	12
Chlorine	Cl	35.5
Chromium	Cr	52
Cobalt	Co	59
Copper	Cu	64
Helium	He	4
Hydrogen	H	1
Iodine	I	127
Iron	Fe	56
Lead	Pb	207
Lithium	Li	7
Magnesium	Mg	24
Nitrogen	N	14
Oxygen	O	16
Phosphorus	P	31
Potassium	K	39
Sodium	Na	23
Sulphur	S	32
Tin	Sn	119
Zinc	Zn	65

Experiment 7.7 The formula of magnesium oxide.

Use a clean crucible and lid. Heat it gently. Let it cool. Measure its mass. Clean a 30 cm strip of magnesium ribbon. Fold it loosely in the crucible. Measure the mass again.

Stand the crucible on a pipe-clay triangle. Heat it gently, then strongly. Lift the lid with tongs. Put it back before smoke escapes. Do this until all the metal has burnt. Let it cool. Measure the mass again.

mass = 15.52g   mass = 16.72g   mass = 17.52g

How many grams of magnesium were used?
How much oxygen combined with it?
How much magnesium would 1 g of oxygen need?
How much would combine with 16 g of oxygen?

From the results shown in the drawing:
16 g of oxygen combine with 24 g of magnesium.
16 g of oxygen is a huge number of atoms. Call this number L. The 16 g of oxygen contain L atoms. The RAM column shows O = 16, Mg = 24. An Mg atom has $1\frac{1}{2}$ times the mass of an O atom. So 24 g of magnesium ($1\frac{1}{2} \times 16$ g) must contain L atoms too. If L atoms of magnesium join with L atoms of oxygen then 1 atom joins to 1 atom. The formula is MgO.

118

This last idea is very important. Try it again. Pretend that we can pick up atoms. Put an oxygen atom on one balance. Put a magnesium atom on another. Its mass is $1\frac{1}{2}$ times as big as an O atom.

Put on more oxygen atoms. Add enough to make 16 g. Count them. It will be a huge number. Call this number L. L is called **Avogadro's Constant Number.**

Put L atoms of magnesium on the other balance. Their mass will be $1\frac{1}{2}$ times 16 g, or 24 g. But our experiment proved that :

   24 g of magnesium combine with 16 g of oxygen
   So L atoms   combine with L atoms
So 1 atom combines with 1 atom. The formula is MgO.

Try another experiment:

Expriment 7.8 The formula of tin oxide

Gently heat a clean dry test tube. Let it cool. Measure its mass. Put in it about 2 g of tin foil. Measure the mass of the test tube and tin. Support the tube in a fume cupboard. Add drops of concentrated nitric acid. When the reaction has almost stopped, add more drops. Go on until no tin is left. Gently boil the liquid away. Heat the solid. Stop when it gives no more brown gas. Let it cool.

Find the mass of tube and solid. The solid is tin oxide. Work out its formula. Use your own results or the ones below. Sn = 119; O = 16.

My results : Mass of tube and tin oxide = 28.17 g
               Mass of tube and tin = 27.63 g
               Mass of empty test tube = 25.62 g
What is the mass of tin used?
What mass of oxygen does it combine with?
How much oxygen combines with 119 g of tin?

Summary

The smallest piece of an element is an atom. It is shown on paper by a symbol. This is the first letter of the name of the element. The Latin name is sometimes used. A second letter is added if needed. The symbol for sulphur is S. Silicon also begins with S. Its symbol is Si. Sodium has the symbol Na. It comes from the Latin name, natrium.

The hydrogen atom is the smallest. Any other atom has a bigger mass. The number of times bigger is called its relative atomic mass. If H = 1, then C = 12, S = 32, Fe = 56, Pb = 207 and so on. Fe and Pb come from the Latin names for iron and lead.

1 g of hydrogen, 12 g of carbon and 56 g of iron contain the same number of atoms. This number is Avogadro's Constant Number, L. We use it to work out the formula of a compound by experiment.

Questions

1. Fill in the blanks as you copy this out.
   An atom is the smallest piece of an _____.
   An atom is shown on paper by a _____.
   The symbol for an atom of nitrogen is _____.
   The symbol for nickel must have an extra _____.
   S is the symbol for one atom of _____.
   S = 32 shows its relative _____ _____.
   The mass of L atoms of sulphur is _____ grams.

2. Use the list on page 113. Which atom in the list has
   a) the highest relative atomic mass?
   b) twice the mass of an oxygen atom?
   c) four times the mass of an oxygen atom?
   d) twice the mass of a calcium atom?
   What is the relative mass of
   e) a lead atom + an oxygen atom?
   f) two oxygen atoms + a carbon atom?

3. The formula of sulphuric acid is H_2SO_4. Write down what this tells us about the acid. Use the words symbol, atom, formula and molecule. Say what each of these words means.

The mole

We have come across many chemical reactions. In each of them new substances are formed. We can find out what these products are. Then we write a word equation. This sums up the reaction.

In Experiment 7.7 we burnt magnesium. It reacts with oxygen. The product is magnesium oxide.

magnesium + oxygen → magnesium oxide

An equation must be based on fact. Our burning experiment shows that:

24 g of magnesium combine with 16 g of oxygen

The magnesium consists of atoms. 24 g of it contain a huge number. We called this number L. Oxygen contains atoms. We proved that 16 g of it is L atoms. L is **Avogadro's Constant Number.**

So L magnesium atoms react with L oxygen atoms.
One magnesium atom joins on to one oxygen atom.
The formula of magnesium oxide is MgO.

We need a name for L atoms of an element. We call 100 of anything a century. 12 is a dozen.
L atoms or molecules is called a mole.

16 g of oxygen is a mole of oxygen atoms. 24 g of magnesium is a mole of magnesium atoms. It is much simpler to say

1 mole of Mg atoms reacts with 1 mole of O atoms.
The formula of magnesium oxide is MgO.

A mole is about 602 300 000 000 000 000 000 000 particles. In words, 602 thousand, three hundred million, million, million. Just call it a mole!

12 is a dozen, 20 is a score,
Avogadro's Number, L, is many millions more.
Are you counting particles? When you call the roll.
Just remember L of them is always called a mole.

We know the formula of magnesium oxide. We can write a formula equation:

$$Mg + O \rightarrow MgO$$

This equation uses an oxygen atom, O. But oxygen gas contains no single atoms. They are joined together in pairs. Each pair is a molecule. It is a **diatomic** molecule. Its formula is written O_2.

This alters the equation. An oxygen molecule will need 2 Mg atoms. Magnesium is not diatomic. It is made up of separate atoms. So we write 2Mg - just as we write 2 eggs or 2p. Each combines with an oxygen atom. Each forms MgO. There are 2 MgO.

The equation is $2Mg + O_2 \rightarrow 2MgO$

Gas elements exist as molecules. Many of them are diatomic. Hydrogen and nitrogen are. Their formulae show this. They are written H_2 and N_2.

If O = 16, then O_2 = 32. An oxygen molecule has 32 times the mass of a hydrogen atom. 32 is its **relative molecular mass,** or its **formula mass.**

32 g of oxygen gas contain L molecules.
32 g is a mole of oxygen molecules.

Did you work out the formula of tin oxide?
My results were : Mass of tin = 2.01 g.
                        Mass of oxygen = 0.54 g.

2.01 g of tin combine with 0.54 g of oxygen

1 g of tin will combine with $\frac{0.54}{2.01}$ g = 0.269 g of oxygen

119 g of tin combine with 119 × 0.269 = 32 g of oxygen

1 mole of tin atoms : 2 moles of oxygen atoms
The formula of tin oxide is SnO_2.

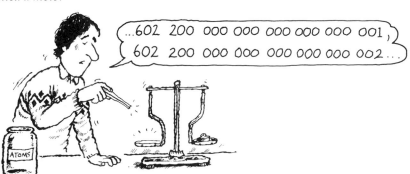

...602 200 000 000 000 000 000 001,
602 200 000 000 000 000 000 002...

Building word equations

Experiment 7.9 Word equations for reactions

1. Put a little chalk into a test tube. Add dilute nitric acid with a dropper. Test any gas given off (page 83). Boil the liquid gently away.

2. Add dilute sulphuric acid to copper turnings.

3. Put dilute sulphuric acid into a test tube. Add black copper oxide. Gently warm it. Add oxide until no more dissolves. Filter. Evaporate.

4. Put lead nitrate solution in a test tube. Add drops of potassium iodide solution. Filter.

Which mixture shows no sign of reaction?
Name any products you know.
Write word equations, as far as you can.

Copper and dilute sulphuric acid do not react.

Chalk and acid give carbon dioxide. Boiling away the liquid leaves a solid. We can write

$$\text{calcium carbonate} + \text{nitric acid} \rightarrow \text{carbon dioxide} + \text{calcium X} + \text{?}$$

Copper oxide and acid react. The solution is blue. Evaporation leaves copper sulphate crystals.

$$\text{copper oxide} + \text{sulphuric acid} \rightarrow \text{copper sulphate} + \text{?}$$

In 4, a yellow solid forms. It is insoluble. Any insoluble substance formed in solution is called a **precipitate**. Filtering removes it. It is shown by an arrow. **Remember that metals can replace metals.**

$$\text{lead nitrate} + \text{potassium iodide} \rightarrow \text{lead iodide} \downarrow + \text{potassium nitrate}$$

Substances which react are called reactants. The new substances formed are the products. Tests usually tell us what they are. Now we can write a word equation. It sums up the reaction.

Tests may not name every product. For example

$$\text{copper oxide} + \text{sulphuric acid} \rightarrow \text{copper sulphate} + \text{X?}$$

How can we find out what X may be?

Every substance has a formula. It can be found by experiment. For copper sulphate it is $CuSO_4$. Copper oxide is CuO. Sulphuric acid is H_2SO_4. Putting in each formula may solve the problem.

$$CuO + H_2SO_4 \rightarrow CuSO_4 + X?$$

Of the reactants we have used Cu and SO_4. O and H_2 are left. They form H_2O, water. The equation is
$$CuO + H_2SO_4 \rightarrow CuSO_4 + H_2O$$

We can also get formula equations by experiment. We need to find out:
   how many moles of each reactant take part
   how many moles of each product are formed
   the formula masses of these substances.

For formula mass use the formula, say H_2SO_4. Symbols show the elements in it. Which ones? Numbers after the symbols show how many atoms.

H_2 means two hydrogen atoms. If $H = 1$, $H_2 = 2$
S means one sulphur atom only. $S = 32$
O_4 means 4 oxygen atoms. If $O = 16$ $O_4 = 64$
The formula mass of sulphuric acid is 98

How many moles of a substance? For a pure solid measure its mass.
$$\text{Number of moles} = \frac{\text{mass}}{\text{formula mass}}$$

We use sulphuric acid as a solution.

One mole of it is 98 g. Dissolve 98 g in pure water. Add more water. Make the volume exactly 1000 cm³. This is called a **molar solution**. It is also called an M solution. We can measure the volume we used. This tells us how many moles.

Formula and equation

Sodium carbonate and hydrochloric acid react. We shall work out an equation for the reaction. We shall use anhydrous sodium carbonate. The acid we use will be a molar solution, or M solution. The formula of hydrochloric acid is HCl.

Experiment 7.10   The equation for a reaction

a) Put sodium carbonate into a test tube. Add the dilute acid. Test any gas given off. Evaporate some of the solution. Examine the crystals formed.

b) Measure 1 g of the anhydrous carbonate. Put it into a conical flask. Add drops of methyl orange indicator. Fill a burette with a molar solution of the acid. Note the reading on the burette.

Run a little acid into the flask. Swirl the mixture around. When reaction is nearly over, add a little more. Add acid until the indicator just changes colour. Note the burette reading at the end.

> What gas is given off?
> What substance are the crystals likely to be?
> What volume of acid reacts with the sodium carbonate?

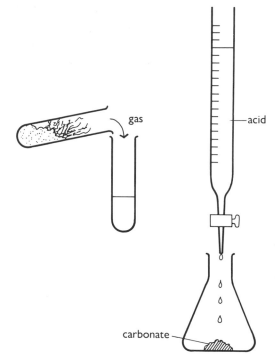

The formula of sodium carbonate is written Na_2CO_3. Na_2CO_3 has 106 times the mass of a hydrogen atom. One mole of it is 106 grams.

The M solution contains 1 mole of acid in 1000 cm^3. Its formula is HCl. We know the volume of acid solution used. This reacts with 1 g of sodium carbonate.

The gas puts out a lighted splint. It makes lime water milky. One product is carbon dioxide. The solution gives cube-shaped crystals. This product is likely to be salt, sodium chloride.

Relative atomic masses
Na = 23, C = 12, O = 16
Na + Na + C + O + O + O
23 + 23 + 12 + 16 + 16 + 16 = 106

The word equation starts as
   sodium carbonate + hydrochloric acid → sodium chloride + carbon dioxide

Suppose 1 g of carbonate reacts with 19 cm^3 of M acid.
106 g of carbonate would need 106 × 19 cm = 2014 cm^3.
One mole of carbonate needs two moles of acid. $Na_2CO_3 + 2HCl$

Na_2CO_3 will give CO_2. Na_2 will form 2NaCl. We still have one O atom and two H atoms. Together these make H_2O. Water is a third product.
The formula equation is
   $Na_2CO_3(s) + 2HCl(aq) \rightarrow 2NaCl(aq) + CO_2(g) + H_2O(l)$

The states of the substances are shown. (s) stands for solid, (aq) stands for 'in water'. What do (g) and (l) mean?

Summary

All matter is made up of atoms. This is called the Atomic Theory. An atom is shown as a symbol. It is usually the first letter of the name of the element. A second letter is added if needed. Some symbols come from Latin names. Sulphur is S, so silicon is Si and silver Ag (Latin name argentum).

The hydrogen atom is the smallest. Any other atom has a bigger mass. The number of times bigger is its relative atomic mass. If $H = 1$, then $C = 12$ and $O = 16$. These are relative atomic masses.

1 g of hydrogen contains L atoms. So do 16 g of oxygen and 12 g of carbon. L is called Avogadro's Constant Number. L atoms are called a mole of atoms.

Atoms combine to form molecules. Many molecules contain two atoms. They are diatomic. Most gas elements are. An oxygen molecule is O_2. If $O = 16$, $O_2 = 32$. 32 is the formula mass of oxygen gas. 32 g of oxygen contain L molecules. It is one mole.

Atoms and molecules are always moving. Those in a solid have fixed positions. They move only slightly. Molecules move freely in a liquid. Some escape through the surface. They pass into the air as vapour. This explains how a liquid evaporates. Molecules of a gas move freely. They spread to fill the vessel holding them. They continually collide.

Heat supplies energy. It makes atoms and molecules move faster. So Atomic Theory can explain all physical changes. It explains chemical change too.

Atoms join together to form molecules. A compound molecule contains different atoms. Its formula shows which ones. The symbols show which elements. Numbers after them show how many atoms of each.

A molecule has a relative molecular mass. It comes from its formula. We shall call it formula mass. Word equations sum up reactions. Formula equations show how many moles take part. The state of each substance can be shown. (s), (l) and (g) mean solid, liquid and gas. (aq) means 'in water'.

A mole is the formula mass of a substance in grams. It is made into 1000 cm^3 of solution. This is called a Molar solution, or an M solution.

Things to do

Repeat Experiment 7.18. Use chalk ($CaCO_3$) and Molar nitric acid. Work out a formula equation. Calcium nitrate is $Ca(NO_3)_2$.

Questions

1. Which of these symbols come from Latin names? sulphur S; zinc Zn; lead Pb; calcium Ca; gold Au; sodium Na; potassium K; tin Sn; silver Ag.

2. Ice, water and steam contain only H_2O molecules. What happens to them when
   a) ice melts,
   b) water evaporates,
   c) water boils,
   d) water freezes?

3. Say what each statement means
   a) the formula of water is H_2O,
   b) $O = 16$,
   c) the relative atomic mass of carbon is 12,
   d) smoke shows Brownian motion.

4. $H = 1$; $O = 16$; $Br = 80$; $C = 12$; $S = 32$; $Fe = 56$; $Mg = 24$; $N = 14$; $Pb = 207$; $Cu = 64$.
   Which of these atoms has the greatest mass?
   Which has four times the mass of a nitrogen atom?
   Which has five times the mass of an oxygen atom?
   Which atom has half the mass of a copper atom?
   How many grams of magnesium make one mole?

5. 1.6 g of copper oxide is heated in hydrogen. Water and 1.28 g of copper are formed. Work out the formula of copper oxide. $Cu = 64$; $O = 16$.

6. Yellow lead oxide is reduced to lead. 2.23 g of it contain 2.07 g of lead. How much oxygen does the oxide contain? Work out its formula. $Pb = 207$

7. The formula of nitric acid is HNO_3. What elements does it contain? How many atoms of each? What is the formula mass of nitric acid ($N = 14$)? What is the mass of one mole of the acid? How many molecules does a mole of acid contain? Explain what is meant by a Molar solution.

Non-metals – carbon

All metals, except mercury, are solids. Their atoms have fixed positions. They are close together. They move very little.

At normal temperatures many non-metals are gases. They exist as molecules. Many of these are diatomic. Each of them contains two atoms. The atoms are bonded together. Hydrogen is like this. Its formula is H_2.

Some non-metals are solids. We shall study two of them, carbon and sulphur. We know a great deal about carbon already.

Carbon

In Experiment 1.14 we heated wood out of air. The solid formed is charcoal. It is one form of carbon.

Experiment 8.1   The properties of charcoal

1. Nearly fill a test tube with water. Add a few drops of litmus solution. Take half of this solution. Add powdered charcoal. Boil the solution gently. Filter it. Compare it with the unused solution. Now use activated charcoal.

2. Dissolve brown sugar in water. Shake half of the solution with activated charcoal. Filter it. Compare it with the unused half.

*3. Fill a test tube with the gas ammonia. Invert it in a trough of mercury. Use a fume cupboard. Push pieces of charcoal into the gas. Push them through the mercury.

Put two drops of bromine in a gas jar. Cover it. Let it fill with vapour. Crush charcoal into small pieces. Put some into the gas jar.

Charcoal is used in gas masks.

4. Find out if charcoal conducts a current. Try to find the mass of 1 cm³ of it (Expt. 4.1).

5. Heat a mixture of black copper oxide and charcoal. Pass any gas into lime water.

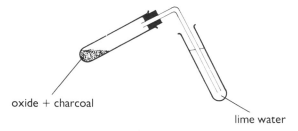

oxide + charcoal

lime water

These properties decide what charcoal is used for. Some of them you have met before. Can you write down some of its uses?

The uses of charcoal

Charcoal does not easily conduct an electric current. Finding the **mass** of a lump of it is easy. Charcoal is made by heating wood. Gases are given off. This leaves millions of tiny holes in the charcoal. They are filled with air. So charcoal is porous. It floats on water. This makes finding the **volume** of a lump difficult.

Mercury rises into the test tube. Some ammonia must have gone. Bromine vapour disappears too. Charcoal takes in, or absorbs, gases and vapours. It does so because it is porous.

Charcoal is used in gas masks. Activated charcoal is best. The person wearing the mask breathes in. The air passes through charcoal. Poison gases and smoke are absorbed. The rest of the air goes through. It has been made pure.

Charcoal absorbs dyes. It takes litmus out of solution. It takes the brown colour out of sugar solution. It is used in making white sugar.

Charcoal can reduce metal oxides to the metal. The arrows show what happens to each reactant. Which arrow shows reduction?

$$2CuO + C \rightarrow 2Cu + CO_2$$

Charcoal was used in early times to make iron. It reduced iron oxide. It was also one of the substances in gunpowder.

Charcoal was made in the forests. Wood was piled up. The pile was covered with turf. Burning wood was dropped through a hole in the top. The hole was covered with turf. Some wood in the pile burnt. It turned the rest into charcoal.

Charcoal is a smokeless fuel. It is still used – in barbecues! Like other forms of carbon, charcoal makes marks on paper.

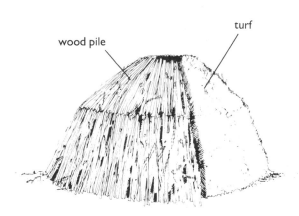

wood pile

turf

Two other forms of carbon

Diamond and graphite are well known forms of carbon. Both are found in nature. Both are made by man.

Graphite

Experiment 8.2 The properties of graphite

1. Rub some graphite on paper. What do you see?

2. Twist the glass stopper in a bottle neck. Rub graphite on the stopper. Twist it in the neck of the bottle again.

3. Find out if graphite can conduct an electric current.

Graphite consists of grey-black crystals. It is soft and greasy to the touch. The crystals look shiny. It is a form of carbon. So it is a non-metal. But its shine is almost like metallic lustre. It is a conductor. It has some of the properties of metals.

Graphite is used as a conductor in electrodes. Carbon is one electrode in dry cells (batteries).

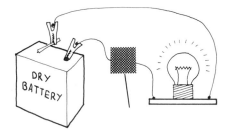

Graphite makes a mark on paper. It is the 'lead' in a pencil. In making leads, graphite is mixed with clay. The mixture is made into a wet paste. It is then forced through holes. The long pieces are baked dry in an oven. They are cased in wood to keep them from breaking.

Making pencil lead.

Graphite is soft. It leaves a trail of crystals on the paper. This is a pencil mark. A pure graphite pencil would be used up quickly. Clay makes it harder. It lasts longer.

H pencils have more clay in the mixture. B pencils have more graphite. They give a blacker mark.

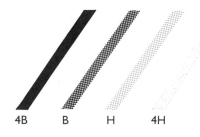

4B   B   H   4H

A stopper twisted in a bottle squeaks. Coated with graphite it moves more smoothly.
Rough surfaces grate over each other. If your bicycle squeaks, two metal parts are touching. Oil is put between them. The film of oil keeps them apart. Oil lubricates the moving parts. Layers of graphite keep surfaces apart. Alone, or mixed with oil, graphite is a lubricant.

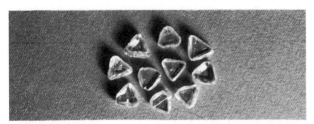

Diamond crystals.

A diamond saw blade cutting hard limestone.

A diamond saw blade cutting fossil footsteps out of rock.

Property	Diamond	Graphite
appearance	transparent	grey-black
mass of 1 cm³	3.5 g	2.2 g
conductor?	no	yes
hardness	very hard	very soft

Diamond

Diamond is a form of carbon. Small pieces of it are called diamonds. They are found in many countries. Most of them come from South Africa. They are found in some kinds of rock. The rock may be part of a long dead volcano.

Some diamonds are clear with no colour. Some are dull and coloured. They all have uses. Uses depend on properties.

Diamond is the hardest known natural substance. It makes scratches on glass. It does this to other hard minerals such as sapphire. Both are used as the stylus (needle) in record players.

Diamond is used in glass cutters. Drills can be studded with diamonds. They are then hard enough to cut through rock. Saws for cutting other hard materials carry diamonds.

Diamond has a definite crystal shape. Its surface reflects light. Light also passes into the transparent crystal. It is reflected inside as well. This is why a diamond sparkles. It is used in jewellery.

Diamond and graphite have many uses. We need far more of both than we find in nature. Graphite is made from sand and coke. They are mixed and heated to 2500 °C. This needs an electric furnace. After some hours, the coke has changed to graphite.

Natural diamonds may have formed in volcanoes. High temperature and pressure might make them. Last century, many people tried to make them. Some cheated. They took natural diamonds and pretended they had made them.

Now graphite is put into a small vessel. It is subjected to huge pressure and temperature. The diamonds formed are small. They are used in drills, glass cutters and so on.

Look at the table. It shows properties of diamond and graphite. They are forms of the same element. Both are carbon.
How can they be so different?

127

The coal industry

Coal was used at least 2000 years ago. It was mined in this country in Roman times.

Where did it come from?

Coal was formed about 240 million years ago. The climate was hot. Areas where rivers ran into the sea were marshy. Giant trees, ferns and rushes grew well there. They died and others grew. This gave layers of rotting vegetation.

Bacteria slowly turned the layers into peat. The rivers covered it with mud and silt. This put pressure on the peat. Water and oxygen compounds were squeezed out. Most of the peat was now carbon. It finally formed coal.

The layers of coal are called seams. They are found at various depths in the earth. Some seams are near the surface. They may even show through it. Then they are called 'outcrops'. They were the first coal to be discovered.

Coal mining began early in Britain, around Newcastle. Records show that it was sent to London by sea. Using it as fuel began air pollution.

By 1700 AD three million tonnes a year were mined. By 1810 it was 10 million tonnes a year. Miners hacked it out of underground seams. They worked as much as fifteen hours a day. So did women and young children.

Mining was dark and dangerous work. Naked flames were the only lighting. Gas came out of the coal seams. The lights made it explode. The gas carbon monoxide was formed. It killed the miners. So did falls of rock and flooding.

Conditions of work were dreadful quite late in this century. By 1913 Britain was mining 287 million tonnes a year. Amounts used are shown:

Year	1925	1950	1960	1970	1975	1984
Million tonnes	176	204	198	150	116	105

A primeval forest.

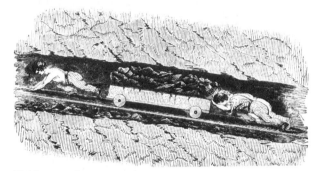
Children working in a coal mine.

Miners in a cage waiting to go underground.

Cutting coal by hand.

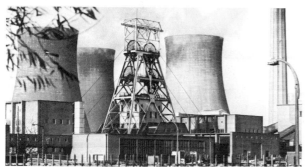
A coal cutting 'shearer' machine at work.

Lea Hall colliery and Rugeley power station.

A gas works producing gas from coal.

A coke oven discharging coke.

The uses of coal

Coal was first burnt as fuel. This wasted useful substances. It put smoke and dust into the air.

In Experiment 1.15 we heated coal out of the air. It gives a gas which burns. It leaves coke. This is a smokeless fuel. Coal tar forms in the tube. So does a watery liquid with ammonia in it. This can be used to make fertilizers.

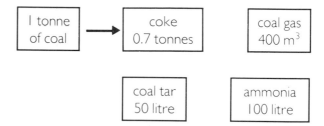

| 1 tonne of coal | → | coke 0.7 tonnes | coal gas 400 m³ |
| coal tar 50 litre | | ammonia 100 litre | |

For many years coal gas was the only gas fuel. About 50% of it was hydrogen, 35% methane and 10% carbon monoxide. This made it poisonous.

The coal tar is collected. Distilling it gives many substances. From them other substances such as dyes and explosives are made. Distillation leaves pitch. This is used in road making.

Coke is carbon. It is used in extracting metals. Other fuel gases are made from coke. Air blown through hot coke makes it burn. This gives carbon monoxide and heat. Then steam is blown through the hot coke.

$$C + H_2O \rightarrow CO + H_2$$

Carbon monoxide and hydrogen are both gas fuels. The mixture is called 'water gas'.

Coal and its products are very important. The rise of the oil industry made it less so. Gas and liquid fuels are easier to transport. They can be piped from place to place. However, the supply of crude oil may run out. Coal as a form of energy is becoming more important.

Allotropy – carbon and sulphur

It is easy to show that graphite contains carbon. We can burn it. A gas is formed. Lime water proves it is carbon dioxide. Graphite must contain carbon.

Sir Humphrey Davy was the first to burn diamond. (He had a rich wife!). This also formed carbon dioxide. Diamond must contain carbon too.

How can we show that they contain *only* carbon? Work out formula masses for O_2 and CO_2. If $C = 12$ and $O = 16$, then $O_2 = 32$ and $CO_2 = 44$. The equation for burning is

$$C + O_2 \rightarrow CO_2$$
$$12 + 32 \rightarrow 44$$

12 g of carbon give 44 g of carbon dioxide.

Now can you answer the question?
Burn 12 g of graphite. Collect all the carbon dioxide it gives. Measure its mass. Is it 44 g? If it is, graphite must be pure carbon only.

Suppose it is not all carbon. Then it contains other substances. The carbon will be less than 12 g. Burning it gives less than 44 g of carbon dioxide.

Diamond can be burnt. 12 g give 44 g of carbon dioxide. Like graphite it contains only carbon atoms. Then why do they differ so much?

Because the carbon atoms are in different patterns.

In **diamond** each atom is joined to four others. Each of these is joined to four more. This pattern goes on in every direction. Each piece of diamond is one huge molecule. It has only carbon atoms.

In **graphite** each atom is joined to three others. Each of these is joined to three more. All these atoms form a flat layer. It has other layers above and below. They are like cards in a pack. Cards slide over each other. So do layers of atoms in graphite.

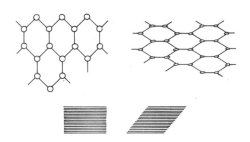

This explains why graphite is soft. If it is rubbed on paper the layers peel off. This leaves a trail of black crystals. We have drawn a pencil line.

The pattern of atoms in diamond makes it hard. The atoms are closer together than in graphite. There are more in 1 cm^3. The mass of 1 cm^3 is larger.

This effect is the **allotropy** of carbon. Allotropy is the existence of an element in different forms. These forms are **allotropes.** They all contain atoms of the element only. They differ in appearance. They have different physical properties.

> *Diamond is hard and glitters,*
> *Graphite's soft as soap.*
> *Odd that each of them should be*
> *A carbon allotrope.*

Many elements have allotropes. Sulphur is one. Sulphur is a mineral. It is found in the earth. It has to be mined. It is sold in two forms. One is a powder. It is called flowers of sulphur. It can also be bought as lumps. This is roll sulphur.

Experiment 8.3 The properties of sulphur

Take a lump of roll sulphur. Find out if it is a conductor. Put it in a mortar. Hit it with a pestle.

Use one piece to find its density. Measure its mass on a balance. Drop into a measuring cylinder of water.

Note how much the water level rises. This rise is its volume.
$$\text{Density} = \frac{\text{mass}}{\text{volume}} \quad \text{in g per cm}^3.$$

Grind some of the sulphur to a fine powder.

What colour is sulphur?
Does it break into pieces easily?
Does it conduct an electric current?
Is its density high or low?
Could you mistake sulphur for gold?

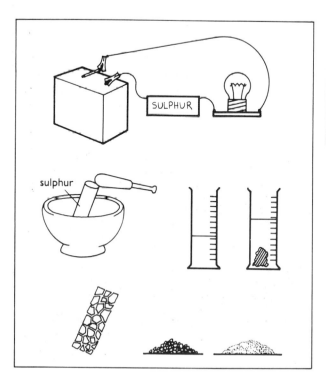

The sulphur breaks easily into small pieces. We ground it into tiny bits. But it cannot be split up into other substances. Sulphur is an element.

Sulphur is a shiny yellow solid. So is gold. Both are elements. But no one mistakes sulphur for gold. Sulphur is brittle. It breaks up when it is hit. Gold is not brittle. When hit, it changes shape.

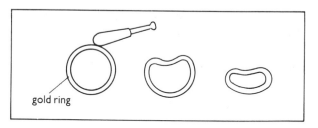

gold ring

Both are shiny. Gold is shiny in a different way. All metals have this shine. It is called metallic lustre. Gold and sulphur differ in appearance.

Gold is a conductor. Sulphur is not. Gold has a much higher density than sulphur. The two differ in what they can do. Their behaviour is different.

Sulphur is a non-metal. Like most solid non-metals:
it is hard but brittle
it is a non-conductor
as compared with metals it has a low density.

Willie's teacher got quite stroppy
When Willie called it allo – tropy.
'The name for crystal forms' said he,
'Must be pronounced al-ot-ro-pee.
But if you mean just one, I hope
You will pronounce it allo-trope.'

Sulphur can appear in many different forms. These are monoclinic sulphur crystals.

More properties of sulphur

Experiment 8.4 The action of heat on sulphur

Use powdered roll sulphur. Hold the test tube in a test tube holder.

1. Fill a test tube two-thirds full of sulphur. Heat it gently for about fifteen seconds. Stop heating. Shake the tube. Heat and shake until all the sulphur melts. Try to avoid any great change in colour.

Fold a filter paper into a cone. Hold it by the rim. Pour in the liquid sulphur. Blow gently on its top. When it becomes solid, slowly open the cone. Do not let the liquid escape. Remember it is very hot.

2. Take a second test tube. Half fill it with sulphur. Melt it gently as before. Now go on heating it. Stop when the liquid is red-brown in colour. Turn the test tube slowly upside down.

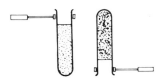

Turn the tube right way up. Heat it until the liquid boils. Fill a beaker with cold water. Pour the boiling liquid into it. Pour away the water. Pick up the solid. Hold both ends and pull. Leave it to stand. Try the pulling test again later.

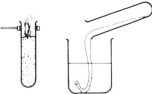

Does sulphur melt easily?
Is its melting point likely to be high or low?
What do you see on the filter paper?
What happens when liquid sulphur is heated?
What happens when the tube is upside down?
Describe the solid left in the water.
Is it sulphur? How can we find out?
Is the solid on the filter paper sulphur?

Sulphur melts to a yellow-orange liquid. It needs only gentle heating. Its melting point is likely to be low. Most non-metals have lower melting points than metals. We can check this by experiment.

Experiment 8.5 The melting point of sulphur

Take a 12 cm length of glass tubing. Heat the middle of it gently. Rotate it in the flame. As the glass gets soft, take it out. Draw it out to a narrow tube.

Cut off a 3 cm length of this. Seal one end in the flame. Let it cool. Jab the open end into powdered sulphur. Two-thirds fill the tube with powder.

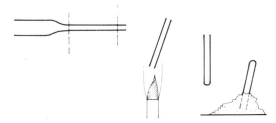

Tie it to a 200°C thermometer. Have the sulphur close to the bulb. Put medicinal paraffin into a small beaker or large test tube. Support the thermometer in it. Gently warm the paraffin. Stir it steadily. Keep the temperature rising slowly.

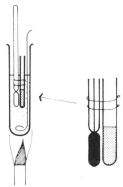

Note the temperature when the sulphur melts. Take out the thermometer. The sulphur becomes solid. Return the thermometer. Find the melting point again.

What is the melting point the first time?
Is boiling water hot enough to melt sulphur?
Is the second boiling point different?

Sulphur melts at about 113°C. Water boils at 100°C. This is not hot enough to melt sulphur.

The liquid sulphur is cooled. It turned to solid. We found its melting point again. It was higher. It may be as high as 119°C. Why is it different?

In the filter paper liquid sulphur cools. It becomes solid. It appears as yellow crystals. They are thin and spiky like needles. They look like sulphur. It is this sulphur in the melting point tube. It melts at 119°C.

We heated liquid sulphur. It gets darker in colour. It becomes orange, red and then brown. We inverted the tube. Nothing ran out. The liquid had become stiff. It was like very thick treacle.

More heating makes it runny again. It is black when it boils. Water cools it suddenly. A soft brown solid is formed. It is called plastic sulphur. Pulling stretches it. It returns slowly to its first length. In time it becomes hard and yellow.

The brown solid and the yellow crystals differ. Are they both sulphur? How can we find out? What else do we know about sulphur? We once burnt it in oxygen. Perhaps we can use this as a test.

Experiment 8.6 Burning sulphur

Put roll sulphur in a combustion spoon. Light it at a bunsen flame. Hold it in a gas jar of oxygen.

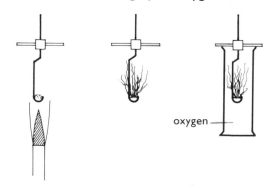

oxygen

Let the sulphur burn. Remove the spoon. Put the burning sulphur in water. Put a lid on the jar. Drop in a dry blue litmus paper. Add water. Shake the jar. Pour in drops of Universal indicator.

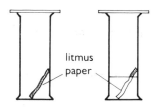

litmus paper

In a second jar burn the spiky crystals. Burn the brown solid in a third jar. Repeat the same tests.

Do they all burn more brightly in oxygen?
Is the flame colour the same?
Did you smell the gas formed?
What do the two indicators show?

All three burn. They burn more brightly in oxygen. They all have the same mauve flame. Each combines with oxygen. Roll sulphur forms an oxide. It has a choking smell. It dissolves in water. The solution contains an acid. The oxide is sulphur dioxide.

The spiky crystals give the same results. So does the brown solid. Both of them must contain sulphur. Can we prove that they contain only sulphur?

The equation for burning is $S + O_2 \rightarrow SO_2$
$S = 32; O = 16.$ So 32   32   64
32 g of sulphur give 64 g of sulphur dioxide.

Take some spiky crystals. Measure their mass. Burn them in oxygen. Collect the sulphur dioxide. Measure its mass. Is it twice the mass of the crystals? Then the crystals are sulphur only.

Sulphur dioxide dissolves in water. The solution contains an acid. It comes from sulphur. It is called sulphurous acid. An equation gives its formula.

$$SO_2 + H_2O \rightarrow H_2SO_3$$

Summary

Sulphur is a mineral. It is found in nature. It is hard but brittle. It is a non-conductor. It has a low melting point. Its density is low. Both are low compared with metals. These are physical properties. Most solid non-metals have these physical properties. Compare them with elements shown in the table.

Element	tin	copper	magnesium	iron	carbon	sulphur
Melting point °C	232	1083	651	1535	3650	113
Density	7.3	8.9	1.7	7.9	2.2	2.1

Its oxide dissolves in water. The solution is acid.

Are there other allotropes of sulphur?

When liquids cool they turn to solids. Some give solid in the form of crystals. Liquid sulphur does. You cooled it in a filter paper cone. Spiky yellow crystals formed. They are shown in the photograph. This shape of crystal is called **monoclinic** sulphur.

In Experiment 1.3 we made salt crystals. A solution of salt in water was evaporated. Can we make sulphur crystals in this way? No! Sulphur does not dissolve in water. We shall try other solvents.

Monoclinic sulphur crystals.

Experiment 8.7 To make sulphur crystals

1. Put 2 cm³ of solvent in a test tube. Add a small amount of powdered sulphur. Stopper the tube. Shake it. If the sulphur dissolves, add more. Put drops of solution on a slide. Look at them with a microscope. Use as solvents a) dimethylbenzene, b) olive oil.

* Your teacher may use carbon disulphide. Work in a fume cupboard. Put the solution in a dish. Cover it with filter paper. Leave it to form crystals.

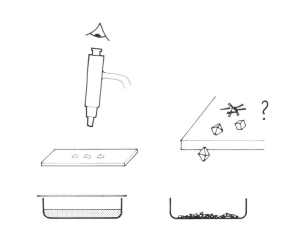

> Do crystals form in all three solutions?
> At what temperature are they formed?
> Do they all have the same shape?
> Does the photograph show this shape?

*2. Take about 20 g of powdered sulphur. Put it in a round bottom flask. Use one with a long neck. Add 60 cm³ of methylbenzene. Heat the flask on a sand tray. When the liquid boils, take its temperature.

Boil it gently for about five minutes. Take its temperature again. Line a beaker with cotton wool. Take a large test tube. Put in 20 cm³ of methylbenzene. Bring it to the boil. Stand it in the cotton wool. Put out all bunsen flames. pour the boiling liquid from the flask into the tube. Look at the liquid a few minutes later.

Rhombic sulphur crystals.

  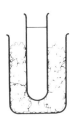

134

Does the boiling point of the liquid change?
When does the sulphur in it melt?
What is the shape of the crystals?
Which photograph on page 134 shows this shape?
At what temperatures are these crystals formed?

In method 1 sulphur dissolves. All three solutions give crystals. The lower photograph shows their shape. They are called **rhombic sulphur** crystals. Rhombic sulphur crystals come from cold solutions.

In method 2 the liquid boils at about 112°C. Some sulphur dissolves. This raises the boiling point. As it rises above 113°C the sulphur melts.

The hot solution is yellow. Cotton wool cuts down heat loss. So the solution cools slowly. This gives time for large crystals to form. They are spiky, like needles. The same kind form in hot liquid sulphur. **Monoclinic sulphur** forms in hot liquids.

Crystals which form above 96°C are monoclinic. Crystals which form below 96°C will be rhombic sulphur. At 96°C one changes into the other. It is called the **transition temperature.** Look again at the photographs. Allotropes differ in appearance.

Rhombic crystals are like pyramids, base to base.

In Experiment 8.5 we melted rhombic sulphur. Its melting point is 113°C. We let it cool. At 113°C it forms monoclinic crystals. Their melting point is 119°C. Allotropes differ in physical properties.

We poured boiling sulphur into cold water. It forms a soft brown solid. This is **plastic sulphur.** It stretches like elastic. In time it hardens again.

Burning shows that the allotropes contain sulphur. But they show differences. These may be caused by impurities. We can show that they are sulphur only. We worked out a method on page 133.

Why are allotropes different?

Sulphur atoms join up in rings. Each ring has 8 atoms. They form the molecule S_8. Rhombic sulphur is made up of these rings. They are packed in a regular pattern. This makes all rhombic crystals alike. They have a regular shape.

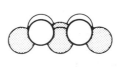

A ring seen from above.   A ring seen from the side.

Above 96°C the packing pattern changes. This gives a new crystal shape. It is monoclinic sulphur. A new packing gives different physical properties. We heated solid sulphur. Heat gives energy. The rings move faster. They break out of the pattern. The sulphur melts. The rings move freely in the liquid. We went on heating the liquid.

More heat gives even faster movement. The rings break up to form chains. These join giving longer ones. These get tangled up like strings of beads. The liquid becomes stiff. It is like thick treacle.

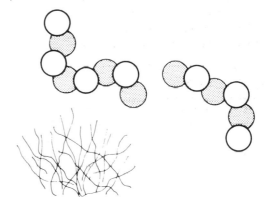

More heating gives more energy. The chains break up into smaller ones. The liquid is runny again.

Water cools the liquid suddenly. The sulphur chains lose energy. They stop moving. They are left in a tangled mess. This is soft plastic sulphur.

Pulling it stretches the tangled chains. Left alone, they go slowly back. The solid is like weak elastic.

Where does sulphur come from?

Sulphur occurs 'native'. This means that the element itself is found in nature. Most of the world's sulphur comes from the USA. It is found in the southern states. People drilling for oil found it by accident. It was thirty years before it was mined.

The sulphur is in a layer of limestone. This is about 150 metres down underground. The problem is quicksand. There is a layer of it between the surface and the sulphur. This prevents normal mining.

In 1890 a man named Frasch solved the problem. Under normal pressure water boils at 100°C. It is not hot enough to melt sulphur. An increase in pressure will raise the boiling point.

Frasch heated water at fifteen times normal pressure. Water is still liquid at 170°C. It is called superheated water. It can melt sulphur.

A hole is bored down to the limestone. Three steel pipes are fed into it. They are placed one inside the other. Superheated water is forced down the outer pipe. It spreads through the limestone. At 170°C it melts the sulphur.

Compressed air is blown down the inner pipe. The liquid sulphur is forced up the middle one. It is kept hot by the superheated water going down. The liquid runs out into huge tanks.

At 170°C only the sulphur melts. Other substances are left behind. So the sulphur which reaches the surface is pure. It cools and turns to a solid.

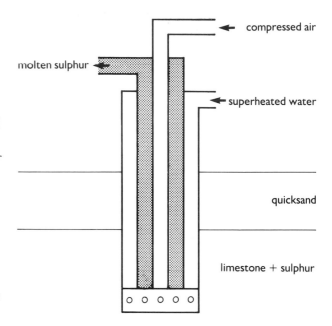

What makes sulphur important?

Huge amounts of sulphur are used to manufacture sulphuric acid. We shall deal with this acid soon. Sulphur is also used to make other sulphur compounds. It is used in making matches and fireworks.

Rubber comes from the sap of the rubber tree. It can also be made from petroleum gases. In both cases it is soft. Sulphur added to the rubber makes it harder and more elastic. This is called vulcanization.

The Frasch process is used to mine sulphur. The photograph shows liquid sulphur pouring into a tank where it will become solid.

Summary

We and the world are made of 'stuff'. This 'stuff' is called matter. It can be solid or liquid or gas. Solid, liquid and gas are the three states of matter.

A particular kind of matter is a substance. Gold is a substance. So is sulphur. So is sulphur dioxide. We recognize a substance by its properties. These are what it looks like and what it does.

An element is a substance. It cannot be split up into other substances. Most elements fall into one of two classes. They are either metals or non-metals. Sulphur is a non-metal. Gold is a metal.

Elements combine. When sulphur burns it combines with the element oxygen. A new substance is formed called sulphur dioxide. Sulphur dioxide is a compound. It can be split up into other substances.

In some changes new substances are formed. They are chemical changes or chemical reactions. The reactions of a substance are its chemical properties.

Melting is a physical change. No new substance is formed. Molten sulphur is still sulphur. So its melting point is a physical property of sulphur.

Some elements exist in more than one form. Each form is the pure element. Different crystal forms of an element are allotropes. They have the same chemical properties. Their physical properties differ. Rhombic and monoclinic crystals are allotropes of sulphur. Allotropy is the existence of allotropes.

A clever young fellow named Frasch
Mined sulphur to make some hard cash.
Steel pipes through the quicksand
Took an air-water mix and
Molten sulphur came up in a flash.

Questions

1. Copy this out, filling in the blanks.

   Sulphur is found in the earth. It occurs _____. Most of the world's sulphur comes from _____. It is in a layer of _____. This is about _____ metres down. A layer of _____ prevents ordinary mining of sulphur.

   Sulphur melts at _____ °C. It is not melted by _____ water. It needs _____ water. This is water heated under greater _____. It is still liquid at _____ °C.

   Label the drawing of the Frasch method. Explain why the sulphur it gives is very pure.

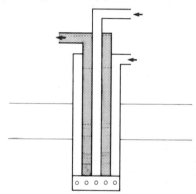

2. Explain what each of the following is. Use two short sentences for each.
   a) an element,
   b) a compound,
   c) an allotrope,
   d) the transition temperature of sulphur,
   e) a chemical reaction,
   f) a physical property.

3. Elements fall into two main classes. Name them.
   In which class would you put
   a) sulphur,
   b) gold?

   In what ways do gold and sulphur differ in
   c) behaviour,
   d) appearance. What are appearance and behaviour called?

   Write word equations for
   e) burning sulphur in oxygen,
   f) dissolving sulphur dioxide in water.

   What colour is litmus in the solution?
   What could be used in place of litmus?

Sulphur dioxide and sulphuric acid

Experiment 9.1 The properties of sulphur dioxide

Fill five jars with the gas. Use a sulphur dioxide syphon. This is better than burning sulphur in the jars. Shake water in each jar. Add a neutral litmus.

Jar 1: Put in coloured flower petals.
Jar 2: Add purple potassium manganate (VII) solution.
Jar 3: Add orange potassium dichromate (VI) solution.
Jar 4: Add hydrogen peroxide solution.
Jar 5: Add barium chloride solution.

> Do the petals change colour?
> What happens to the manganate (VII) solution?
> What happens to the dichromate (VI)?
> What does barium chloride produce?

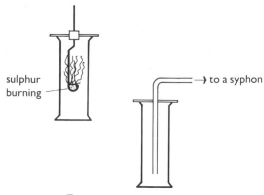

VII means seven, 7
VI means six, 6.

sulphur dioxide + water $\rightarrow$ sulphurous acid
$SO_2(g)$ + $H_2O(l)$ $\rightarrow$ $H_2SO_3(aq)$

(g) stands for gas
(l) stands for liquid.
(aq) means 'in water'

Each gas jar contains sulphurous acid solution.

Some flower petals lose colour. They are bleached. Sulphur dioxide bleaches wood, straw and wool.

Purple manganate (VII) loses its colour. The orange dichromate (VI) becomes green. These are likely to be new substances. A chemical change has occurred. Then sulphurous acid must have changed. The litmus is still red. Is the acid sulphuric acid, H_2SO_4?

Jar 5 had sulphurous acid solution. Barium chloride reacted with it. A white solid was formed. The solution became 'milky'. A solid formed in solution is a precipitate. This one is white barium sulphite.

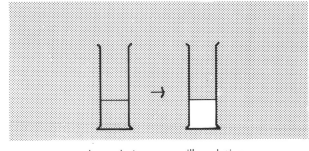

clear solution   milky solution

Experiment 9.2 Sulphurous acid or sulphuric acid?

Add barium chloride solution to gas jar 4. Compare it with jar 5. Add dilute hydrochloric acid to both.

Both form white precipitates. The one in jar 5 dissolves in the acid. The one in jar 4 does not.

Substances added to jars 1 to 4 lose oxygen. They form new substances. These are different in colour.

Sulphurous acid gains this oxygen. It is turned into sulphuric acid.

H_2SO_3 + oxygen $\rightarrow H_2SO_4$

We put hydrogen peroxide in jar 4. Its formula is H_2O_2. It loses oxygen. It becomes H_2O, water. Loss of oxygen is reduction. The peroxide is reduced.

Sulphurous acid gains the oxygen. It is oxidized to sulphuric acid. Barium chloride shows the change.

$$H_2SO_3(aq) + H_2O_2(aq) \rightarrow H_2SO_4(aq) + H_2O(l)$$

With barium chloride solution:
sulphurous acid gives   a white precipitate which dissolves in hydrochloric acid
sulphuric acid gives   a white precipitate which does *not* dissolve in the acid

The manufacture of sulphuric acid

This acid is very important. Huge amounts of it are used. It was first made by heating iron sulphate ore. Later, burning sulphur was used.

It was burnt in a spoon. The spoon was pushed into a large bottle. Water in the bottle dissolved the sulphur dioxide. Oxygen from air oxidized this sulphurous acid. It gave sulphuric acid solution.

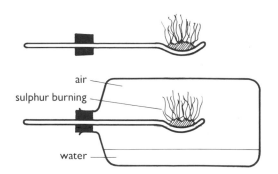

air
sulphur burning
water

The process was very slow. So a substance giving oxygen was mixed with the sulphur. Oxidation was speeded up. Sulphuric acid was made much faster.

Uses for it increased. So new ideas improved the method. Large rooms replaced the bottles. They were lined with lead. This is not attacked by the acid. Nitrogen oxides supplied the oxygen. This method is the 'Lead Chamber Process'. It gives the acid *solution*. We need a way to make pure acid.

The newer method also uses oxidation. The gas is oxidized, not its solution. SO_2 in water gives the acid H_2SO_3. What substance in water would give the acid H_2SO_4? What is its formula and name?

$$SO_2(g) + H_2O(l) \rightarrow H_2SO_3(aq)$$
$$?\quad + H_2O \rightarrow H_2SO_4(aq)$$

The answer is sulphur trioxide, SO_3. Sulphur dioxide and oxygen may combine to make it. They do not react easily. If they did, sulphur dioxide and air would react. How do we speed up reactions?

1. We can use heat.
2. We can use a catalyst.

Summary

Sulphur is a solid non-metal. It exists in more than one form. Each form is an allotrope. Each contains sulphur atoms only. So allotropes have the same chemical properties.

In allotropes atoms are packed in diffrent ways. So their physical properties differ. They have different shapes, densities and melting points. One allotrope can change into another.

Sulphur allotropes burn. The main product is the gas sulphur dioxide. It has a choking smell. It dissolves in water. The result is a solution of sulphurous acid. It takes oxygen from substances. Orange potassium dichromate (VI) becomes green. Other coloured substances are bleached white. The oxygen converts sulphurous acid to sulphuric acid.

Questions

1. What is meant by the allotropy of sulphur? Describe how rhombic sulphur crystals are made. How could you prove that they contain sulphur?

2. Wood pulp is used to make paper. It is first bleached white with sulphur dioxide. Explain this bleaching. The paper slowly turns brown in air. Explain this reverse change.

3. Some fuels contain traces of sulphur. What is put into the air by burning them? Will this gas dissolve in rain? Will the rain affect litmus? The rain falls. What else may it affect?

4. You are given a solution. How would you show that it contains an acid? It gives a white precipitate with barium chloride solution. Name two acids which can do this. How do you show which one?

Sulphur trioxide and sulphuric acid

*Experiment 9.3   To make sulphur trioxide

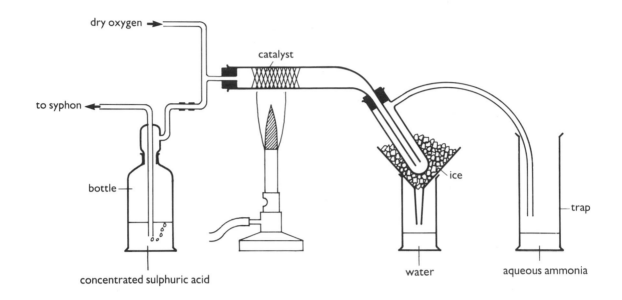

The drawing shows a suitable apparatus. Set it up in a fume cupboard. Use oxygen from a cylinder. Use sulphur dioxide from a syphon. Bubble each gas through concentrated sulphuric acid. This dries them. It also shows how fast they pass through.

Cool the test tube with ice. Put aqueous ammonia into the gas jar. This traps unused gases. The catalyst is fibre coated with platinum. Heat it fairly strongly. Check that the bottles are connected correctly. Start the gases bubbling through. Oxygen should be at one bubble per second. Sulphur dioxide should be twice as fast.

What does the product look like?

Hold a little of it on a spatula in air. Drop small amounts of it into water in a test tube. Divide this solution into four parts.

Add litmus to one part.
Add Universal indicator to a second part.
Add barium chloride solution to the third. Then pour in dilute hydrochloric acid.
Put a strip of magnesium in the fourth portion. Test any gas given off with a lighted splint.

What do the tests tell us about the solution?

Reaction happens in the catalyst tube. Work out word and formula equations for it.

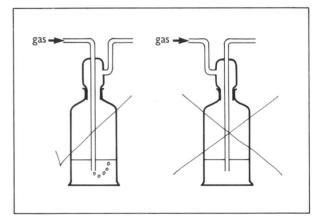

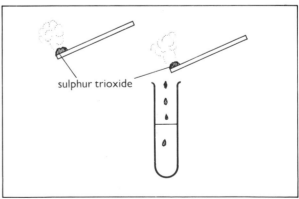

White crystals form in the cooled tube. They are sulphur trioxide crystals. In the warmer air the solid crystals evaporate. This forms smoky fumes.

sulphur trioxide

In water they make a hissing noise. They give smoky fumes. Sulphur trioxide does not easily dissolve. It does react. The solution is acid.

This acid reacts with barium chloride. A white precipitate forms. It does not dissolve in dilute hydrochloric acid. The solution contains sulphuric acid, H_2SO_4.

The word equation for the preparation is

Sulphur dioxide + oxygen → sulphur trioxide

$SO_2 \ldots SO_3$ needs one O atom. Oxygen is diatomic. One O molecule will oxidize two SO molecules. The formula equations will be:

$$2SO_2\,(g) + O_2\,(g) \rightarrow 2SO_3\,(g)$$
$$SO_3\,(s) + H_2O\,(l) \rightarrow H_2SO_4\,(aq)$$

The manufacture of sulphuric acid

Sulphur or a sulphur ore is burnt in a stream of air. It uses up oxygen. It forms sulphur dioxide. Excess air is used. Then there is still some oxygen left. It reacts to give sulphur trioxide.

reactants

The mixed gases are freed from dust. They are cooled to 450 °C. They are compressed to three times air pressure.

conditions

The gas mixture passes over the hot catalyst. Vanadium(V) oxide is used. It is cheaper than platinum.

catalyst

Sulphur trioxide is formed. It does not react smoothly with water, so it is passed into concentrated sulphuric acid. The liquid formed is called oleum, $H_2S_2O_7$.

making the acid

$$SO_3 + H_2SO_4 \rightarrow H_2S_2O_7$$

oleum

With water it gives sulphuric acid.
$$H_2S_2O_7 + H_2O \rightarrow 2H_2SO_4$$

This is the Contact Process. The gases and catalyst come into contact. They react. The reaction gives heat. It is exothermic. It is also reversible. So reaction conditions are important. They are chosen to give as much sulphur trioxide as possible. Most sulphuric acid is made in this way. It gives purer acid.

The Contact Process

A catalyst is used. It alters the speed of reaction. It usually speeds it up. It is still there at the end. Some catalysts slow reaction down. Using a catalyst is called **catalysis.**

141

Concentrated sulphuric acid

corrosive

Pure sulphuric acid has no colour. It is a dense, oily liquid. Treat it with great care. Keep it off your clothes, your skin and the bench. Wear eyeshields. If you have to wash it off, use a large amount of water.

*Experiment 9.4   The properties of sulphuric acid

1. Half fill a 100 cm³ beaker with water. Take its temperature. Pour in a little concentrated acid. Stir the mixture with a glass rod. Take the temperature again.

2. Fill a 100 cm³ beaker one third full with the acid. Stick a label on it. Mark the acid level on the label. Let it stand in air for a few days. Note the level again.

3. Put crystals of blue copper sulphate in a beaker. Add concentrated acid. Stir with a glass rod. Note any change. Pour off the acid into another beaker. Add drops of water to the solid left behind.

4. Dip a paper spill into the acid. Take it out. Hold it over the sink.

5. Put a layer of sugar in a tall 100 cm³ beaker. Make it damp by adding drops of water. Pour in some concentrated acid.

What happens to the temperature in test 1?
Why does the volume increase in test 2?
What happens to the copper sulphate?
How does water affect the white solid?
What is the black substance in tests 4 and 5?

The mixture of water and acid becomes very hot. It may even boil with enough acid. After a few days the volume in test 2 is bigger. The acid takes water vapour from the air. Sulphuric acid has a 'liking' for water.

Blue copper sulphate crystals contain water. It is water of crystallization. The formula is written $CuSO_4.5H_2O$. Concentrated sulphuric acid removes it. White anhydrous copper sulphate is formed. Adding water gives the blue form again. This is a test for water.

Sugar turns into black carbon. It rises out of the beaker. Smoke and steam pour out of it.

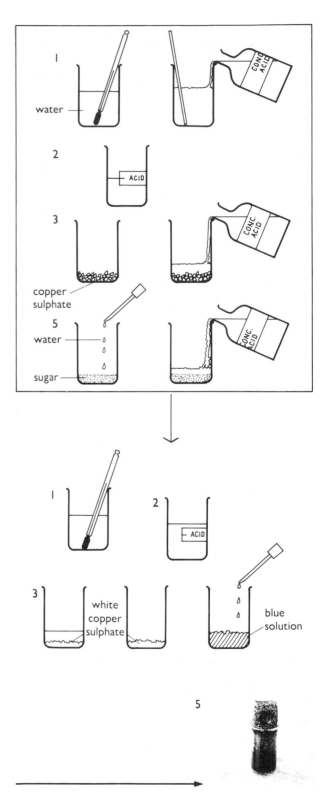

A molecule of sugar is $C_{12}H_{22}O_{11}$. There is no water in it. It does contain hydrogen and oxygen. It has twice as many hydrogen atoms as oxygen. So has water, H_2O. The acid takes away this hydrogen and oxygen as though they were water. Carbon remains.

$$C_{12}H_{22}O_{11} + \text{sulphuric acid} \rightarrow 12C + 11H_2O$$

Some water turns into steam. This blows the carbon out of the beaker. The acid does the same thing to paper. It can also do it to skin, clothes and wood. Heat is also produced.

Taking water from a substance is dehydration. Some compounds contain two H atoms for every O atom. Taking away hydrogen and oxygen is also dehydration. Concentrated sulphuric acid dehydrates.

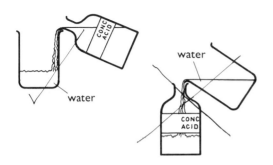

Never add water to concentrated sulphuric acid. Heat is formed. It turns the first water added to steam. This may blow acid into the air. If acid is to be diluted, add acid to water. Wear eyeshields.

Summary

Sulphur dioxide is a gas. Pressure turns it into a liquid. This can be stored in a syphon. The gas dissolves in water. The solution contains sulphurous acid.

$$SO_2(g) + H_2O(l) \rightarrow H_2SO_3(aq)$$

This acid takes oxygen from other substances. Removal of oxygen is reduction. Sulphuric acid is formed. Adding oxygen to a substance is oxidation.

$$H_2SO_3(aq) + \text{oxygen} \rightarrow H_2SO_4(aq)$$

Purple potassium manganate(VII) solution gives oxygen. It needs sulphuric acid as well. The products are colourless. The colour change shows that oxidation is happening.

Sulphuric acid is manufactured by the Contact Process. Sulphur dioxide and air are mixed. They pass over a catalyst. It is vanadium(V) oxide at 450°C. Sulphur trioxide is formed. It is dissolved in concentrated sulphuric acid. Oleum, $H_2S_2O_7$, is formed. It can be diluted with water to give the pure acid.

$$2SO_2(g) + O_2(g) \rightarrow 2SO_3(g)$$
$$SO_3(s) + H_2SO_4(l) \rightarrow H_2S_2O_7(l)$$

Sulphuric acid takes up water. It becomes dilute. It takes water from the air. It removes water of crystallization. It takes hydrogen and oxygen from substances. These are examples of dehydration. Heat is formed. This makes adding water to concentrated acid dangerous. Always add acid to water.

$$CuSO_4.5H_2O \rightarrow CuSO_4 + 5H_2O$$
$$C_{12}H_{22}O_{11} \rightarrow 12C + 11H_2O$$

Questions

1. What is dehydration?
   Explain how concentrated sulphuric acid dehydrates
   a) blue copper sulphate crystals,
   b) sugar.
   Are the changes exothermic?

2. Sulphuric acid is made by the Contact Process. Explain two ways in which sulphur dioxide is made for this Process.
   What is sulphur dioxide mixed with?
   What catalyst is used?
   What temperature is it at?
   The sulphur trioxide formed is dissolved in concentrated acid.
   What substance is formed?
   Why is the oxide not dissolved in water?

3. The two oxides of sulphur are acidic. Write an equation for the reaction of each of them with water. Name the two acids. What test is used to show which is which? Give one use of sulphur dioxide.

4. Write an equation for each of these changes
   a) sulphur burning in air,
   b the reaction between sulphur dioxide and oxygen,
   c) sulphur trioxide dissolving in sulphuric acid.

143

Dilute sulphuric acid and sulphates

Many acids are like sulphuric acid. they come from non-metal oxides. These oxides react with water.

Water contains hydrogen. So does the acid. Hydrogen is what makes it an acid. So hydrogen is put first in the formula. The acid contains the non-metal. This says which acid it is. So it comes next in the formula. Other elements come last.

The formula tells us that:

$$H_2SO_4$$

it is an acid   it contains sulphur   and oxygen too

Magnesium reacts with acid. What happens to it?

Experiment 9.5 The action of acid on metal

Put magnesium in a test tube. Add dilute sulphuric acid. Test any gas with a lighted splint. Add more magnesium. When no more dissolves, filter it.

Evaporate the filtrate. Boil most of the water away. Let it stand. Put drops of hot solution on a slide. Look at it through a microscope.

  The mixture effervesces. What gas is it?
  What happens to the magnesium?
  Is it reacting with water or acid?
  Did you notice any other change?
  What does the hot solution give?
  Write an equation for the reaction.

The gas is hydrogen. The magnesium disappears. It must be in solution. The reaction is fast. Water and magnesium react very slowly. The magnesium must be reacting with acid. The solution is very hot.

The reaction gives heat. It is an exothermic reaction. When it stops there is still magnesium there. This shows that all the acid has gone.

The word equation is :
  magnesium + sulphuric acid $\rightarrow$ hydrogen + ?

Put in each symbol and formula :
  $Mg\,(s) + H_2SO_4\,(aq) \rightarrow H_2\,(g) + ?\,(aq)$

This seems right. Metals can displace hydrogen. We used this to set up the Activity Series. The gas is diatomic. Its formula is H_2. Magnesium is an active metal. So the likely equation is :
  $Mg\,(s) + H_2SO_4 \rightarrow MgSO_4\,(aq) + H_2\,(g)$

The SO_4 part of sulphuric acid is called **sulphate.** $MgSO_4$ is the formula of **magnesium sulphate.** It is a **salt.** It belongs to the same family as common salt. Crystals of it appear in the hot solution.

A salt is a substance formed by replacing the hydrogen of an acid by a metal.

Can we make copper sulphate by this method? No! Copper is low in the Activity Series. It is below hydrogen. It cannot push hydrogen out of an acid. We know copper oxide and acid react.

Experiment 9.6 To make copper sulphate

Take a 100 cm³ beaker. Fill it a quarter full with dilute sulphuric acid. Add black copper oxide. Gently heat it. Stir with a glass rod. If all the oxide dissolves, add more. Do this until some oxide is left in the liquid.

Let it cool. Filter it into an evaporating dish. Gently heat the dish. Stop heating when half the water has gone. Put the solution in a flat dish. Let it stand.

Do the oxide and acid give a gas?
Why do we use too much oxide in the acid?
Pick a well-shaped crystal from the dish.
Draw it in your book. Write the equation.

Copper oxide dissolves. No gas is formed. A blue solution forms in the beaker. Some oxide was left. This shows that no acid is left. The solution is acid-free.

It passes through the filter. Crystals form in the hot solution. They have a definite shape. They are pure. Their full name is copper (II) sulphate. For (II) say 2.

Copper oxide is a **base**. A base reacts with an acid. The products are a **salt** and **water** only.

Experiment 9.7 A third method for salts

Use a 100 cm³ beaker. One-quarter fill it with dilute sulphuric acid. Add a little malachite. Wait until reaction stops. Add more malachite if none is left.

When no more dissolves, filter. Evaporate off about half the liquid. Leave it to form crystals.

Make magnesium sulphate crystals by this method. Use magnesium carbonate instead of malachite.

What gas is formed in both cases?
Work out an equation for each reaction.

Malachite is copper carbonate, $CuCO_3$. The formula of magnesium carbonate is $MgCO_3$.

The gas given off is carbon dioxide. Copper sulphate is formed from malachite. If some remains, all the acid is gone. Its hydrogen has been replaced by a metal. It has been neutralized. A salt is formed.

A copper atom replaces two hydrogen atoms. So does a magnesium atom. Both join SO_4. SO_4 is sulphate.

So $CuCO_3 + H_2SO_4 \rightarrow CuSO_4 + CO_2 + ?$
O from CO_3 and H_2 are left. They form H_2O, water.

$$CuCO_3 + H_2SO_4 \rightarrow CuSO_4 + CO_2 + H_2O$$

> *Willie, finding life a bore,*
> *Drank some H_2SO_4.*
> *Willie's teacher saw that he*
> *Was filled with $MgCO_3$.*
> *Now he's neutralized, its true,*
> *But he's full of CO_2.*

What is wrong with the verse?

Acid and alkali

An alkali is a base. It is a base which dissolves in water. It reacts with an acid. The products are a salt and water.

Experiment 9.8 To make sodium sulphate

1. Fill a burette with dilute sulphuric acid. Hold it over a sink. Run out some acid. This brings the acid level on to the scale. It also fills the part below the tap. Support the burette in a stand. Note the level of the acid.

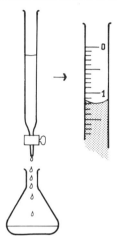

Take a pipette bulb. Practice with it using water. Then pick up 25 cm³ of sodium hydroxide solution. Put this into a conical flask. Add three drops of methyl orange indicator.

Run acid into the flask. Swirl the mixture. Add more acid. Stop when the colour shows signs of change. Now add acid in drops. Swirl the flask after each drop. Stop when the colour just turns pink. Take the reading on the burette.

The acid and alkali react. The solution contains a salt. It is neutral. At most it has one drop of acid too much.

How can we get the salt from the solution?
What colour will the crystals be?

We can evaporate the solution. However, the indicator is still there. So the crystals will be pink. The indicator has done its job. It has told us how much acid is needed.

2. Take a clean flask. Put in 25 cm³ of alkali solution. Do not add an indicator. Take the burette reading. Run in the same volume of acid as before. Evaporate this solution.

A copper atom can replace two hydrogen atoms. A sodium atom can replace only one. H_2SO_4 has two hydrogen atoms. It needs two sodium atoms to replace them. Sodium hydroxide is NaOH. The equation will be

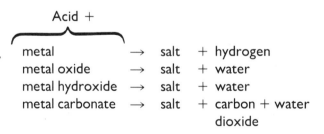

$$2NaOH + H_2SO_4 \rightarrow Na_2SO_4 + 2H_2O$$

Any acid contains hydrogen. The rest of the acid is called the **anion**. Some common ones are shown. A salt is named from its metal and anion.

Acid	Anion
carbonic acid	carbonate CO_3
hydrochloric acid	chloride Cl
nitric acid	nitrate NO_3
phosphoric acid	phosphate PO_4
sulphuric acid	sulphate SO_4
sulphurous acid	sulphite SO_3

We have four methods of making a salt. Each uses an acid. Its hydrogen is replaced by a metal. Four kinds of substance can do this.

Acid +

metal	$\rightarrow$	salt	+ hydrogen
metal oxide	$\rightarrow$	salt	+ water
metal hydroxide	$\rightarrow$	salt	+ water
metal carbonate	$\rightarrow$	salt	+ carbon + water dioxide

Metal oxide and hydroxide are bases. A base and an acid react. They form a salt and water only.

> *Willie's sister's teacher taught her,*
> *Acid + base gives salt and water.*

Insoluble salts

There is a fifth method of making salts.

Experiment 9.9 To make insoluble salts

1. Put sodium sulphate solution in a test tube. Add drops of barium chloride solution. Filter.

> What do you see?
> Is barium chloride a salt?
> Does it contain a metal and an anion?

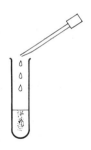

We mixed two salt solutions. We know metals can replace metals. Let them change places. The word equation for the change will be

$$\text{sodium sulphate} + \text{barium chloride} \rightarrow \text{sodium chloride} + \text{barium sulphate} \downarrow$$

Notice the arrow!

Two salts can react and form two new salts. They do so only if one new salt is insoluble. An insoluble substance formed in solution is a **precipitate**. It appears in the solution. We can see it. Which of the two new salts is it?

We shorten precipitate to ppt
Just say the letters p p t

One new salt is sodium chloride. It is common salt. We know it dissolves in water. It is not insoluble. So the white ppt is barium sulphate.

2. Add barium chloride solution to other salt solutions. Use sodium carbonate, sodium nitrate, sodium sulphite, zinc sulphate, magnesium sulphate. Pour into each tube dilute hydrochloric acid.

> Is a ppt formed in every tube?
> Are all the ppts white?
> Which ppts do not dissolve in the acid?

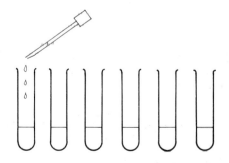

Only sodium nitrate gives no ppt. All the others give a white ppt. Each ppt is a barium salt. Only barium sulphate does not dissolve in acid. So we have a test for sulphates.

$$\text{X sulphate} + \text{barium chloride} \rightarrow \text{X chloride} + \text{barium sulphate}$$

The white ppt forms. It does not dissolve in dilute hydrochloric acid. Other ppts dissolve.

Questions

1. Sodium hydroxide is an alkali. What does this mean? It can react with (a) nitric acid, (b) sulphuric acid. Name the salts formed. Describe how you would make one of these salts.

2. What is a salt? They are made from acids. What four types of substance are added to an acid? Which would you use to make (a) copper sulphate, (b) magnesium sulphate? How would you know if all the acid had been used up?

3. Complete this as you write it in your book. All acids contain _____. The formula of an acid stands for one _____ of it. An acid has the formula HPO_3. One molecule of it contains _____ hydrogen atom, one atom of _____ and _____ oxygen atoms. The formula of its sodium salt is _____. Its magnesium salt is $Mg(PO_3)$ ___.

The uses of sulphuric acid

Nearly four million tonnes of sulphuric acid are made in Britain every year.

1. It is used in making sulphates. Many of these important salts are used in industry. *Farmers* and other *food producers* use them too.

For example, aluminium sulphate is used in making *paper*. It is needed to make *dyes* cling to cloth. It is also used in *sewage* treatment. Plants such as *potatoes* and *vines* suffer from blight. This is a fungus disease. Bordeaux mixture kills or prevents blight. It is a solution of copper sulphate mixed with lime. Copper sulphate also prevents *timber* rotting.

Sulphuric acid is used to make fertilizers. This is a sulphate of ammonia silo.

2. The acid is used in making *fertilizers*. One of these is ammonium sulphate (sulphate of ammonia). Another is 'superphosphate'. Making these uses about one third of the acid manufactured.

3. Sulphuric acid is used in making many other substances. These include *man-made fibres* such as rayon. This is used in clothes, carpets and tyres. The acid is used in making *soap*. With plant oils the acid forms *detergents*.

Explosives, dyes and many medical *drugs* need the acid in their manufacture. It is used in refining crude oil, or *petroleum*.

The acid is also used in making explosives.

4. Many *metals* corrode in air. A layer of oxide forms on them. To prevent this the metal may be painted. It can also be plated with a layer of less active metal. The layer of oxide must first be taken off. It is dissolved in acid. This is called 'pickling'. The clean surface is washed free of acid before *plating*.

5. About one sixth of the acid made is used in the *paint* industry. Barium sulphate and titanium dioxide are solids made for paints.

6. About eighteen million *cars* and *lorries* use the roads in Britain. Most of them run on a *'wet battery'*. The liquid in it is sulphuric acid. So millions of litres of sulphuric acid travel on Britain's roads every day.

Cars use batteries containing sulphuric acid.

Look again at the words in italic on the opposite page. They show how much we depend on the sulphuric acid. The more prosperous the country the more acid it uses.

A wet battery contains fairly concentrated acid. It is made up of a number of cells. Each cell consists of two plates, or poles. They dip into the acid. The negative pole of one cell is joined to the positive pole of the next.

A dry battery is also a series of cells. But each cell has a moist paste instead of a liquid. Electric current passes through paste and acid. Both must be conductors.

Wet batteries.

Dry batteries.

Experiment 9.10 Is sulphuric acid a conductor?

Set up the apparatus using a dry battery. Any source giving six volts will do. Connect its positive pole to a switch. Connect this to a carbon rod. Connect the negative pole to a bulb. Connect this to a carbon rod. Switch on. Touch one carbon rod with the other.

Take three 100 cm³ beakers. In one put a little pure sulphuric acid. In the next put dilute sulphuric acid and in the third, distilled water.

* 1. Hold the rods in the pure acid as far apart as possible. Switch on the current. Move the rods closer together. Do not let them touch. Take them out of the acid. Wash them several times with distilled water.

2. Dip the rods in the dilute acid. Repeat the same tests as in the pure acid. Again wash the rods with distilled water. Dip them in the distilled water and do the same tests as before.

When the carbon rods touch in air the bulb lights up. Why does this happen?

The rods are then separated by three different liquids. Which of the three liquids can conduct a current?

In these tests we have used a very small voltage.

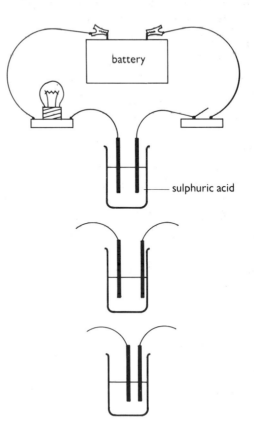

149

Electrolysis

A battery supplies electric current. It flows from one pole of the battery to the other. It must have something to flow through. The poles must be joined by a conductor. A current always produces heat.

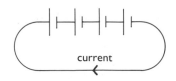
current

We connected the poles to carbon rods. A bulb was included. It has a thin wire, or filament. A current makes the filament very hot. It glows. The bulb lights up. This proves that a current is flowing.

In the 'OFF' position a switch has an air gap. Switching on puts a conductor across the gap. With air between the rods the bulb does not light. Current is not flowing. Air must be a non-conductor. If the rods touch, current passes. The bulb lights.

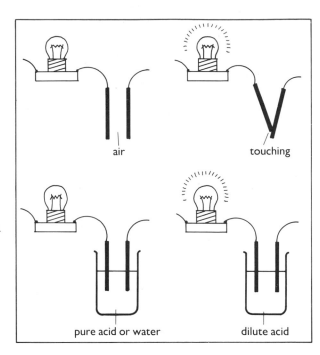
air

touching

pure acid or water

dilute acid

We dipped the rods in a liquid. The liquid is between them. If the bulb lights, the liquid must be a conductor. The bulb does not light with pure acid. Distilled water gives the same result. Both are poor conductors. With the rods in dilute acid the bulb lights. Dilute acid is a conductor.

Can you see why we washed the carbon rods? Did you notice any other result with dilute acid?

Current passes into the dilute acid through one rod. It leaves through the other. The rods are called **electrodes.** Bubbles of gas form on both rods. They start when the current is switched on. They stop when it is switched off. The gases formed are new substances.

The change is called **electrolysis** (el-ec-trol-iss-iss). It is the passage of a current through a liquid, forming new substances.

The liquid is called the **electrolyte**. The electrode joined to the positive pole is called the **anode**. The one joined to the negative pole is the **cathode**. The new substances appear at the **electrodes**.

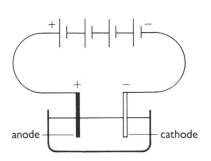
anode

cathode

High voltages make air, water and other substances become conductors. Lightning is a current in air.

Experiment 10.1 To find out what the gases are

Set up the apparatus. Fill the vessel with dilute sulphuric acid. Fill two test tubes with the acid. Invert them in the vessel. Hold one over each electrode. Switch on. Collect the gases. Test both with a lighted splint. Try any other test you need.

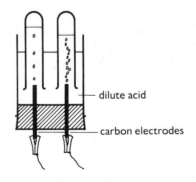

dilute acid

carbon electrodes

What gas is formed at each electrode?
Which of the gases has the bigger volume?
Can you guess how much bigger its volume is?

Experiment 10.2 To compare the volumes of the gases

The apparatus is called a voltameter. Pour dilute sulphuric acid into it. Open the taps on the side tubes. Let the acid fill these tubes. Close the two taps.

The electrodes are pieces of platinum foil. Each is welded to a platinum wire. This is sealed into glass tubing. The tubing passes through the stopper.

Switch on. Switch off when the gas formed is enough to test. Hold a test tube over each tap. Open the taps. Let the side tubes fill up. Close each test tube with your finger. Test the gases in them.

Switch on again. When one side tube is nearly full of gas, switch off. Read off the volume of each gas.

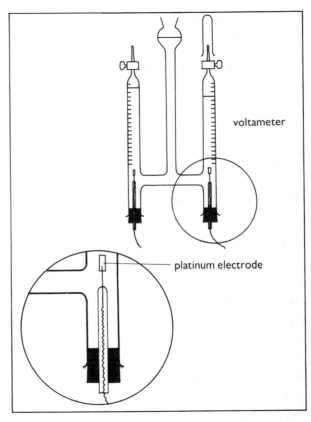

voltameter

platinum electrode

Why does a voltameter have a centre tube?
What are the two gases which form at the electrodes?
How do their volumes compare?
Which gas appears at the cathode?
Where do these gases come from?

At the anode (+): oxygen appears. A splint burns brighter in it. It relights a glowing splint.

At the cathode (−): hydrogen forms. Its volume is twice the volume of oxygen. It gives a squeaky pop with a lighted splint.

Summary

Water is a poor conductor. So is pure sulphuric acid. The dilute acid is a mixture of the two. It is a good conductor. It can be electrolysed.

At the cathode: hydrogen is set free.
At the anode: oxygen is set free.
The volume of hydrogen is twice the volume of oxygen.

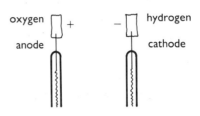

oxygen + − hydrogen
anode   cathode

Our results so far leave questions to be answered.

How does electrolysis work?

Three questions to answer are:
1. Where do the gases come from?
2. Why does hydrogen appear at the cathode, not the anode?
3. Why do we get a good conductor by mixing two poor conductors?

We shall try to answer question 1 first.

Can the gases come out of the electrodes?
Carbon and platinum are elements. Elements cannot be split up into other substances. Neither can provide hydrogen or oxygen. Yet both do this in electrolysis.

Do the electrodes react with the acid?
The gases could be products of reaction. If so, the electrodes would be used up. Anode and cathode are the same element. It could not give *different* gases by reacting with the *same* acid. Gases appear *when the current is switched on.* They stop *as soon as it is switched off.*

Gas molecules must reach the electrodes from somewhere. Testing other liquids may give helpful results.

Experiment 10.3 Electrolysis of other liquids

Use the apparatus of Experiment 4.1. Put a bulb in the circuit. Collect and test any gas. Look for any other changes. Check the electrodes for changes.

Use ethanol (meths), dilute hydrochloric acid, salt solution, oil, copper(II) sulphate solution, zinc sulphate solution and *mercury.

Write your results in the form of a table.

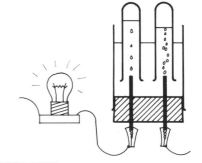

Liquid used	Conductor or not	Result at the cathode	Result at the anode	Any other result
copper(II) sulphate solution	conductor	a red-brown solid forms	oxygen appears	the blue solution becomes paler

Ethanol and oil do not conduct. The bulb does not light up.
Mercury is a metal. It conducts. No new substances are formed.
The others are conductors. In each of them new substances form.

The substances set free – how much?

Metals and hydrogen form at the cathode. Non-metals are the products at the anode.

Take the results for copper(II)sulphate solution. The blue solution becomes paler. Copper sulphate is being used up. A red-brown solid forms on the cathode. This must be copper. It can only have come from the copper(II)sulphate electrolyte.

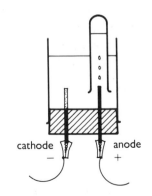

cathode — anode +

The products of electrolysis come from the electrolyte.

What makes hydrogen and copper move to the cathode? And how much?

*Experiment 10.4 What masses are formed?

Almost fill a 100 cm^3 beaker with 0.5 M copper(II) sulphate solution. Almost fill another with 0.05 M silver nitrate solution. Clean two copper plates with fine emery paper. Clean silver electrodes.

Wash the metals with water. Let them drain. Wash them with meths. Let them drain. Dip them in propanone. Take them out and let them dry. Measure the mass of each electrode.

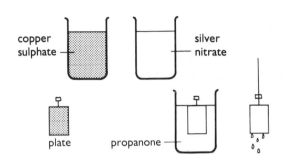

Support the copper electrodes in the copper sulphate solution. Connect them up as shown. Note which of them is the anode. Support the silver electrodes in the silver nitrate solution. Label the cathode.

The milliammeter is to measure the current. Adjust the rheostat to get a current of 0.1 ampere. Start a stop clock. Use the rheostat to keep the current constant. Let it pass for about one hour.

Switch off. Measure the time accurately. Take out each electrode with care. Wash them with distilled water, meths and propanone as before. Let them dry. Measure the mass of each of them again.

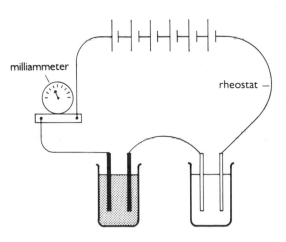

What happens to the copper cathode?
Does the mass of the silver cathode change?
What happens to the two anodes?
Does the copper(II) sulphate solution become paler?

Each cathode gains in mass. Copper is deposited on the copper cathode. The silver cathode gains silver.

The mass of each anode gets less. They are both losing metal. The results show that

gain in mass of cathode = loss in mass of anode

The copper(II)sulphate does not become paler. We know it loses copper to the cathode. It must gain copper from the anode. Gain and loss just balance.

153

Ions and electrolysis

We have answered question 1.
The electrolyte provides the products of electrolysis.
How do they get to the electrodes? The answer is ions.

We know that   electrical charges of the same sign repel
unlike charges attract each other.

+ repels +
− repels −
+ attracts −

Copper is deposited on the cathode. Then copper particles must move towards it. The cathode is negatively charged. The copper particles must be positively charged. Negative attracts positive.

These particles are called copper **ions**. Each ion is a copper atom with two units of positive charge. It is written Cu^{++} or Cu^{2+}. The 2 explains why they are copper(II) compounds.

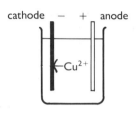

cathode   −   +   anode

$\leftarrow Cu^{2+}$

At the cathode the ions lose their charge. They become neutral copper atoms. They join on to the cathode. A layer of copper metal is formed on it.

Copper(II) sulphate, $CuSO_4$ has no over-all charge. If the copper ion is Cu^{2+}, SO_4 must have an equal negative charge. It is the sulphate ion, SO_4^{2-}.

Cu^{2+} in $CuSO_4$?
Then SO_4^{2-} is the other ion.

The positive anode attracts negative ions . They are called **anions**. Positive ions move to the cathode. They are called **cations**. The current is moving ions.

Two changes are possible at the anode.
Metal ions may leave the anode.
Anions may form non-metals there.

$SO_4^{2-} \rightarrow$

$Cu^{2+} \leftarrow$

A liquid conducts if it contains ions. The moving ions are the current. Water is a poor conductor because it has few ions. Substances such as copper(II)sulphate put ions into the water. Acids provide hydrogen ions.
They are atoms with one unit of positive charge. We write them H^+. Alkalis give OH^- ions.

acid gives H^+ ions
alkali gives OH^- ions

Hydrogen ions move to the cathode. They lose their charge. They become neutral atoms. These join up in pairs to form molecules. Hydrogen gas is set free.

Salt crystals are made up of ions. They are Na^+ and Cl^-. The crystals do not conduct. The ions are there but they cannot move freely. When salt dissolves in water the ions separate. They can move freely in the solution. The salt becomes a conductor. Electrolysis happens.

H^+
$\leftarrow$
H^+

You will learn later why metals and hydrogen have positive ions.

Heat melts crystals. Ions may be able to move in the liquid. If so, the liquid may conduct. We can test this using lead iodide. Its ions are Pb^{2+} and $2I^-$.

*Experiment 10.5 Electrolysis of a molten salt

Connect a battery through a bulb to two carbon rods. Put lead iodide crystals into a large test tube. Dip the carbon electrodes into the crystals. Switch on the current. Heat the crystals gently.

Does lead iodide conduct?
When?
What are the products at cathode and anode?

The bulb lights up when the salt melts. The liquid is a conductor. It must contain moving ions. Beads of shiny silvery lead form at the cathode. Violet vapour appears at the anode. The vapour is iodine.

Lead iodide is chosen because it melts easily. Its melting point is low. Other salts have high melting points.

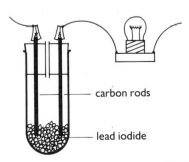

carbon rods

lead iodide

Quantities in electrolysis

Here is a set of results for Experiment 10.4:
Mass of copper deposited on the cathode = 0.120 g
Mass of silver on the silver cathode = 0.405 g
Current passing through both solutions = 0.10 ampere
The time for which the current passed = 3600 seconds

Current and time can both be altered. Experiment shows that
1. doubling the current doubles the mass of product,
2. doubling the time also doubles the mass of product formed,
3. mass of product is proportional to current × time.

$$\begin{array}{ccc} \text{current} & \times & \text{time} & = & \text{quantity of electricity} \\ \text{(ampere)} & & \text{(second)} & & \text{(coulomb)} \end{array}$$

$0.1 \times 3600 = 360$ coulombs

The mass of silver is more than the mass of copper.
Why? Silver ions have a bigger mass than copper ions.
A silver ion Ag^+ has one unit of charge.
A copper ion Cu^{2+} has twice this charge.
It needs twice as many silver ions to give the same quantity of electricity.

$Cu = \text{\textcircled{64}}$; $Ag = \text{\textcircled{108}}$

0.405 g of silver are given by 360 coulombs
108 g would need $\dfrac{360 \times 108}{0.405} = 96\,000$ coulombs

$96\,000$ coulombs = 1 faraday

108 g of silver = L ions or L atoms (one mole)
A mole of silver atoms needs 1 faraday. A mole of copper atoms needs 2 faradays.
   Cu^{2+} 2 faradays per mole set free
   Ag^+ 1 faraday per mole set free

155

The uses of electrolysis

Read through this whole section again. What substances are set free at the anode and cathode? Do the results suggest uses for electrolysis?

1. New substances form at the electrodes
Many substances are manufactured by this method. Electrolysis of salt solution (brine) produces three. They are sodium hydroxide, chlorine and hydrogen. You will learn of other substances later.

2. A layer of metal may form on the cathode
This is used to coat an object with metal. The object could be a cheap metal spoon. The spoon is used as the cathode. The coating metal is the anode. It could be silver. The solution contains silver ions. They move towards the cathode. A silver layer is deposited on it. We now have a 'silver' spoon.

The method is called electroplating. The coating layer is often used to protect a metal from corrosion.

Experiment 10.4 used copper electrodes. They stood in copper(II) sulphate solution. Copper formed on the cathode. Copper was lost from the anode.

This method is used to purify copper. The impure copper is the anode. The cathode is a thin sheet of pure copper. Copper leaves the anode. It is deposited on the cathode. Other metals in the anode are left behind.

Sea water is very corrosive. The upright post has been eaten away, but the railings, which are coated with zinc (galvanized), have not rusted at all.

Electrolysis is used to purify copper.

3. Electrolysis of molten salts gives metals
Compounds of active metals are difficult to split up. These metals can be made by electrolysis. Active metals react with water. So a molten salt is used.

Sodium, calcium, magnesium and aluminium are made like this. The anode products are useful too.

Things to do

1. Wash a new nail with detergent. Rinse in water. Use the nail as a cathode. Use a zinc anode. Put them in zinc sulphate solution. Make this acid with dilute sulphuric acid. Electrolyse the solution.

2. Put a copper anode in dilute copper sulphate solution. Use in turn as a cathode
   a) a plastic spoon,
   b) a strip of stiff paper.
   Dry them. Rub them all over with a 4B pencil. Try them again.
   What is deposited on each cathode?

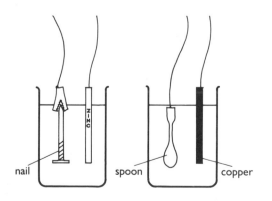

nail   spoon   copper

Summary

Dilute sulphuric acid is a typical acid. It reacts with metals. A salt and hydrogen are formed.

$$Mg(s) + H_2SO_4(aq) \rightarrow MgSO_4(aq) + H_2(g)$$

It dissolves metal oxides. A salt and water form.

$$CuO(s) + H_2SO_4(aq) \rightarrow CuSO_4(aq) + H_2O(l)$$

Carbonates in acid effervesce. The gas given off is carbon dioxide. This is a third method for making a salt.

$$MgCO_3(s) + H_2SO_4(aq) \rightarrow MgSO_4(aq) + CO_2(g) + H_2O(l)$$

Acids react with alkalis. They neutralize each other. The products are a salt and water. A salt is made by replacing the hydrogen of an acid by a metal. In all four methods the salt is in solution.

Pure sulphuric acid is a poor conductor. So is water. Mixing them gives dilute acid. This is a good conductor. We shall find out how it forms ions.

Electrolysis means passing a current through a liquid to form new substances. Conductors dip into the liquid. They are called electrodes. The negative electrode is the cathode. The positive electrode is the anode. The liquid is called the electrolyte.

Ions move through the electrolyte. An ion is an electrically charged atom. Some ions are charged groups of atoms. Positive ions move to the cathode. They are called cations. Negative ions move to the anode. They are anions.

At the electrodes ions lose their charge. New substances are set free.
The mass of each new substance depends on
a) the current passing
b) the time it passes
c) the relative atomic or formula mass

One mole of ions is set free by
a) one faraday of electricity for one-charge ions
b) two faradays for ions with double charges
c) three faradays for ions such as Al^{3+}

At the cathode, metals and hydrogen are set free. At the anode, non-metals form or metal is lost.

Electrolysis has many uses.
Substances are manufactured by this method. The electrolyte can be a molten salt. Very active metals are extracted by this method. Metals form at the cathode. This can be used to
a) plate one metal with another
b) purify a metal such as copper

Questions

1. Write one sentence about each word to explain what it means
electrode; cathode; electrolyte; anode; electrolysis; positive ion; anion.

2. What is electro-plating?
In what two ways is it useful?
Explain how it can be done.

3. Fill in the gaps as you write this in your book:

Current in an electrolyte is moving _____. Cations move towards the _____. Metals are set free at the _____. So is the gas called _____. Non-metals form at the _____. They have _____ charged ions. The anode may lose _____. The copper ion is written _____. A hydrogen ion is _____.

Inside the atom

In 1896 Becquerel discovered radioactivity. He did it by accident. He was testing substances which glow in the dark. One of them was a uranium ore.

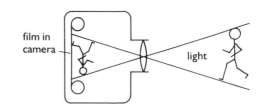

We take photographs using light. We let it strike films or plates. So they must be kept away from light before use. They are wrapped in black paper.

Becquerel had put some wrapped plates in a drawer. One had been affected. Near it was the uranium ore. Tests showed that radiation came from the ore. Like light, it had affected the plate. To do this it had to pass through black paper.

Rutherford showed that the radiation was of three types. He called them alpha, beta and gamma rays. These are the Greek letters a, b and c.

Alpha radiation is positively charged helium ions.
Beta rays are also particles. They are much smaller than hydrogen atoms. They are negatively charged. They are called electrons.
Gamma radiation consists of waves similar to X-rays.

$$\alpha\text{-rays} = He^{2+}$$
$$\beta = electrons$$
$$\gamma = waves$$

So far, atoms have been thought of as solid spheres. Electrons changed this picture. If they exist, atoms must contain smaller particles. How are these arranged in the atom?

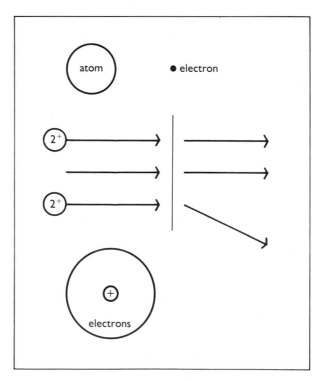

In 1909, Geiger and Marsden set up a stream of α-particles. They aimed them at a thin sheet of gold. Most of them went straight through. They passed through gold atoms. These seemed to be mostly empty space. Atoms were far from being solid.

Some α-particles were deflected. It suggested that they were being repelled. Some part of the atoms must be positively charged. We now believe that:

1. Every atom has a tiny centre. This is called its **nucleus**. It is positively charged.
2. Every atom contains negatively charged **electrons**. These are in groups around the nucleus. The rest of the atom is empty space.
How big is the nucleus?

Imagine that we could magnify a hydrogen atom.
Make it as large as a sports hall.
The nucleus would become as big as a pin head.
An electron would be as big as a speck of dust.
The rest of the atom would be the empty space.
You can see why α-particles pass through gold.

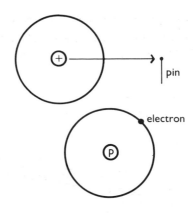

The hydrogen atom is the simplest one. Its relative
atomic mass is 1. It contains one electron. This has
one unit of negative charge. The atom is neutral. Its
nucleus must have one unit of positive charge. It
contains a particle called a **proton.**

An electron has a very small mass. It is about $\frac{1}{2000}$ of
the mass of a hydrogen atom. The rest of this atom
is a proton. Its mass is almost the same as a
hydrogen atom. If H = 1, then a proton = 1.

In 1932, Chadwick discovered a third particle. This
is the **neutron.** It has no charge. Like the proton, its
mass is 1 atomic mass unit.

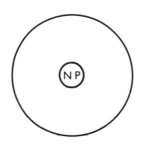

Neutrons and protons are found in the nucleus. The
nucleus contains most of the mass of the atom.

Particle	Mass in atomic mass units	Charge	Where is it in the atom?
proton	1 a.m.u.	+	nucleus
neutron	1 a.m.u.	none	nucleus
electron	$\frac{1}{2000}$ a.m.u.	−	groups around the nucleus

The next simplest atom is the helium atom, He. It has two protons in
its nucleus. So it must have two electrons.
He = 4. The mass of a helium atom is four units. Two protons
provide two units. The nucleus must also contain two neutrons.

nucleus = 2P + 2N
2 electrons

The lithium atom comes next. It has three protons. Li = 7.
How many neutrons? Where will the protons and neutrons be?
How many electrons in this atom?

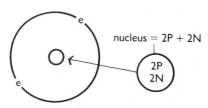

helium atom

In every atom:
1. the number of protons = the number of electrons.
This number is the **atomic number** of the element.

2. The main mass of the atom is in the nucleus. It contains protons and
neutrons. The number of both together is the **atomic mass number**.
Electrons are in groups around the nucleus of the atom.

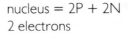

Atoms and ions

Two numbers label the atom of an element. **Atomic number** is the number of protons in the atom. The atom contains the same number of electrons.

Every atom has a relative atomic mass. As a whole number this is called the **mass number.** It is the number of protons and neutrons in the atom.

The elements can be placed in order of atomic number. The table shows the first twelve of them.

Atomic number	1	2	3	4	5	6	7	8	9	10	11	12
Symbol	H	He	Li	Be	B	C	N	O	F	Ne	Na	Mg
Mass number	1	4	7	9	10	12	14	16	19	20	23	24

Consider oxygen, symbol O. Its atomic number is 8. This means that an oxygen atom has eight protons. Then it must have eight electrons. Its mass number is 16. Protons + neutrons = 16. Its nucleus has eight protons and eight neutrons.

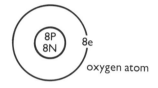
oxygen atom

The symbol for sodium is Na. Its mass number is 23. The nucleus of its atom has twenty-three protons and neutrons. The atomic number of sodium is 11. Its atom has eleven protons. Then it must have 23 minus 11 neutrons, or twelve. Because it has eleven protons it must have eleven electrons.

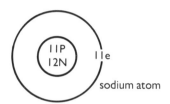
sodium atom

The electrons are in groups. They surround the nucleus. The groups are called **shells.** Moving atoms collide. Their electron shells will collide first.

A helium atom has two electrons. A neon atom, Ne, has ten. Helium and neon are noble gases. They have no simple reactions. This is why they are called noble. Their electron shells must be inactive.

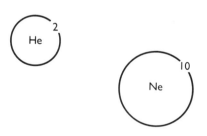

Modern theory says that other atoms become like them. They reach the same shell arrangement as a noble gas atom.
To do this they gain or lose electrons.

An oxygen atom has eight electrons. This is two less than a neon atom. Neon is the nearest noble gas to oxygen. So an oxygen atom gains two electrons.
It now has two extra negative charges. It is an oxygen ion, O^{2-}.

A sodium atom, Na, has eleven electrons. This is one more than a neon atom. The Na atom will lose an electron.
Its nucleus has eleven protons ($11+$). It has ten electrons ($10-$).
It has become the sodium ion, Na^+.

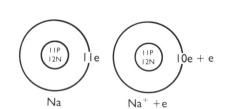
Na $Na^+ + e$

Hydrogen is the first element. Its atom has one electron. It can lose this. It becomes the ion H^+. Hydrogen and metal atoms lose electrons. They then have more protons than electrons. Their ions are positively charged. In electrolysis they move to the negative cathode. It provides electrons. These electrons turn positive ions back into atoms.

Metals and hydrogen are formed at the cathode.

Non-metal atoms gain electrons. They form negative ions. These move to the positive anode. The anode removes electrons. The ions are turned back to atoms.

Non-metals are set free at the anode in electrolysis.

$$metal\ atom \rightarrow metal\ ion + electrons$$
$$+ ion$$

$$non\text{-}metal\ atom + electrons \rightarrow non\text{-}metal\ ion$$
$$- ion$$

Questions

1. Write this in your book. Fill in the blank spaces.

    A radioactive substance gives off _____. There are _____ types of radiation. They are named from the first _____ letters of the _____ alphabet.

    Alpha radiation consists of helium _____, He^{2+}. Beta radiation consists of particles. they are called _____. They have a _____ charge. Their masses are very _____, $\frac{1}{2000}$ of the mass of a _____ atom. Gamma radiation consists of _____ similar to _____ rays.

    Every atom has a small centre called its _____. It contains _____ and _____. A proton has the same mass as a _____ atom. It is _____ charged. The number of protons in an atom is its _____ _____.

    An atom has no overall charge. Its electrons are _____ charged. Its electrons and _____ are equal in number. Electrons in an atom are in groups. They surround the _____. The groups are called _____. The number of protons and neutrons in an atom is its _____ _____.

2. The symbol for magnesium is Mg. Its atomic number is 12. Its mass number is 24. Fill in the blanks.

    A magnesium atom has a central _____. It contains protons and _____. The total number of these is _____. A magnesium atom has _____ times the mass of an H atom. It contains _____ protons. It must have _____ electrons. The neon atom has ten electrons. A magnesium atom will lose _____ electrons. It still has _____ protons but only _____ electrons. It will be _____ charged. It has become the magnesium ion, Mg^{-+}.

3. Read the statements about each particle. Choose the ones which are correct. Write your answer as
    A if only 1, 2 and 3 are correct,
    B if only 1 and 3 are correct,
    C if only 2 and 4 are correct,
    D if only 4 is correct,
    E for any other combination of correct answers.

    a) A proton
        1 is positively charged,
        2 has the same mass as an H atom,
        3 is found in the nucleus,
        4 has the same mass as an electron.
    b) A neutron
        1 is positively charged,
        2 is found in the nucleus,
        3 has no electrical charge,
        4 has the same mass as an H atom.
    c) An electron
        1 is negatively charged,
        2 has the same mass as an H atom,
        3 is contained in a shell in the atom,
        4 is positively charged.

161

Radioactivity

What is it?

Becquerel discovered it by chance. He showed that uranium compounds gave off rays. The effect was named radioactivity. The compounds are radioactive. Other radioactive compounds were discovered. They contain elements with large atoms. These large atoms give the rays.

The rays are of three types. They were named alpha, beta and gamma. These are the Greek letters a, b, c.
α-rays are helium ions, He^+.
β-rays are streams of electrons.
γ-rays are waves like X-rays or light.

alpha is α
beta is β
gamma is γ

What causes it?

A radioactive atom is unstable. Its nucleus breaks up. It does so of its own accord. 'Bits' fly out of it. We say the atom 'spontaneously disintegrates'. The nucleus alters. This is a nuclear change. It is **not** a chemical change.

In chemical changes atoms react. They gain, lose or share electrons. These electrons are in their outside shells. These nucleus of each atom stays the same.

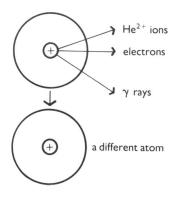

He^{2+} ions
electrons
γ rays

a different atom

What is the result of radioactivity?

Two quantities are enough to label an atom. They are atomic number and mass number. Each can be shown with its symbol. An atom of sodium is written as $^{23}_{11}Na$.

mass number 23
symbol Na
atomic number 11

The nucleus can 'decay' in two ways. In alpha-decay it emits a helium ion, He^{2+}. This has two protons and two neutrons. The mass number of the new substance is four units less. Its atomic number is two less. It becomes another element.

uranium 235 $\rightarrow$ thorium
$^{235}_{92}U \rightarrow \, ^{4}_{2}He + \, ^{231}_{90}Th$

In beta-decay the nucleus loses an electron. It comes from a neutron. It leaves an extra proton in the nucleus. Radium becomes actinium.

neutron $\rightarrow$ proton + electron
$^{228}_{88}Ra \rightarrow \, ^{228}_{89}Ac$ + electron

What does radiation do?

All three types of ray have energy. They pass through paper and metals.
They affect photographic film.
If they strike zinc sulphide they make it glow. They turn gas molecules into ions. In damp gas this leaves a vapour trail.
These effects identify a radioactive substance. They also trace where it is.

Radioactivity is a source of energy. Power stations run on it. It damages or destroys living tissue. It can strike and alter the nuclei of other atoms.

How long does it last?

A radioactive element emits rays. The nuclei of its atoms alter. After a time half of them will have changed. This time is the 'half-life' of the element. It can be a fraction of a second. It can be thousands of years. It is constant for any one element.

half-life 15 years
200 atoms → 100 left unchanged
Then 100 → 50 left unchanged
will also take 15 years
So will 50 → 25 unchanged.

How can it be used?

Radioactive elements are found in nature. They can also be made. Normal compounds are hit by radiation. For example, sodium is not radioactive. A sodium compound is bombarded with neutrons. These change it.

Some sodium nuclei take in a neutron. The atoms are still sodium. The properties of the compound are not changed. But it is radioactive. It is called a radioisotope. It can be traced by the rays it gives.

$$^{23}_{11}Na + \text{neutron } ^{1}_{0}n \rightarrow ^{24}_{11}Na$$

For example, plants take phosphates from the soil. A radioisotope phosphate can be added. Its chemical properties are the same.
The plant takes it from the soil.
It gives out rays. These show us where it goes. We can find out how plants use soil phosphates.

They can locate tumours, cracks in metals, leaks in underground pipes, wear in car engines and many more. Their radiation destroys living cells. This gives them many medical uses. They are used against cancer.

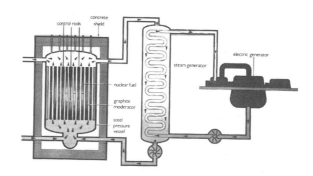

We can use radioactivity to find out how plants deal with phosphates from the soil.

Nuclear Energy

A large nucleus can be split into smaller ones. Uranium 235 splits if hit by neutrons. This is nuclear fission. The fission gives more neutrons. These hit other nuclei. A chain reaction is set up. Huge amounts of energy are released. The change can be explosive. This happened in the first atomic bomb.

Fission can be controlled by controlling the neutrons. Metals such as cadmium 'mop' them up. This slows down fission. Graphite slows neutrons down. It moderates them. This increases the rate of fission. Heat from it is converted to electrical energy. Small atoms can be built into bigger ones. This is nuclear fusion. It was used in the 'hydrogen bomb'.

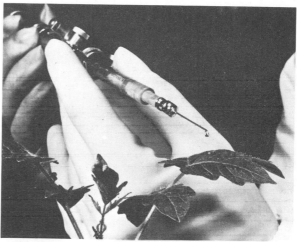

How a nuclear reactor works. Heat from the fuel makes steam which drives a generator.

163

The halogens

We now know many properties of concentrated sulphuric acid. We shall look at one more. Be careful! It attacks skin, paper and wood.

Experiment 11.1 Concentrated sulphuric acid and salt

Wear eyeshields. Put salt, sodium chloride, into a test tube. Use as much as will cover a $\frac{1}{2}$p coin. Add ten drops of the acid. Use a dropper. TAKE CARE!

Note what happens. Hold a dry litmus paper in the tube. Then use a wet litmus paper. Put in a Universal Indicator paper and a lighted splint.

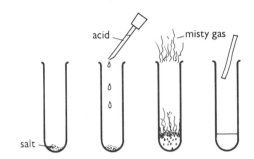

What do you see when the acid is added?
How do you know that a gas is formed?
Can you see the gas inside the tube?
Can you see it outside the tube?
What properties of the gas do the tests show?
Does it have the same effect on dry and wet litmus?

Experiment 11.2 To find out if the gas dissolves in water

Put three or four small pieces of rock salt in a test tube. Add ten drops of concentrated acid as before. Collect the gas in a large test tube. Cover the mouth of the tube. Turn it upside down in water. Add a litmus paper to the water in the tube.

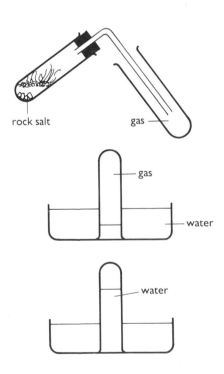

Does rock salt give a faster stream of gas?
What happens in the large test tube in the water?

The mixture of acid and salt froths and bubbles. This is effervescence. It shows that a gas is being given off. It cannot be seen inside the tube. It becomes misty when it gets into the air.

The gas affects litmus. A dry paper changes only slowly. A wet litmus paper turns red at once. Water rises quickly in the large tube. A great deal of gas dissolves. The solution is an acid. The gas does not burn. It puts out a lighted splint.

What element does every acid contain?
Acids react with metals. What is the usual product?
Acids and metals form salts. How did we define a salt?

All acids contain hydrogen. It can be replaced by a metal. The result is a salt. The hydrogen often appears as a gas.

The misty gas dissolves in water. The solution contains an acid. The hydrogen of this acid may come from the gas. It may come from water.

Remember that sulphur trioxide and water react. They form sulphuric acid.

$$SO_3 + H_2O \rightarrow H_2SO_4$$

The hydrogen in this acid comes from water.

*Experiment 11.3 Does the misty gas contain hydrogen?

Use the apparatus shown below. Put lumps of rock salt in the test tube. Add concentrated sulphuric acid. Put iron wool in the tube. Heat it gently. Collect the gas formed. Remove the delivery tube from the water.

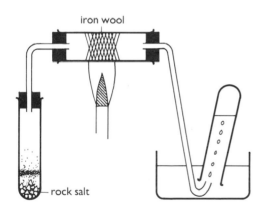

iron wool

rock salt

The rock salt and acid give a steady stream of gas.
What happens to the iron?
What gas is collected over the water?
Does the misty gas contain hydrogen?

Iron is an element. It contains only iron. It reacts with the misty gas. The gas formed is hydrogen. This hydrogen cannot have come from iron. Hydrogen must come from the misty gas. The misty gas is a hydrogen compound.

This compound must contain at least one other element. To find out which one we must remove the hydrogen. Oxygen from other compounds can do this by oxidation.

Experiment 11.4 To oxidize the misty gas solution

Wear eyeshields. Use a concentrated solution of the gas. Put it into three test tubes. Add to one a pinch of copper(II) oxide.
Add a pinch of manganese(IV) oxide to the second.
Drop crystals of potassium manganate(VII) into the third.
If nothing happens warm the tubes.

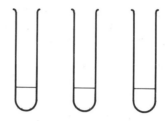

Hold a dry blue litmus paper in the tube. Then use a wet litmus. Try a Universal Indicator paper.

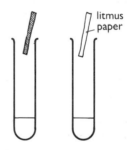

litmus paper

Which of the three mixtures give a gas?
Is the gas coloured?
What is its effect on dry litmus paper?
What happens to the wet litmus paper?
Do the potassium manganate(VII) crystals give a purple colour?

Chlorine and hydrogen chloride

The misty gas in water gives an acid. We oxidized it. We used three substances. All three contain oxygen. Manganese(IV) oxide gives a gas. So does potassium manganate(VII). Copper(II) oxide gives a green solution but no gas.

The gas is pale green. It slowly turns a dry litmus paper red. It finally turns it white. It is being bleached. Wet litmus is bleached at once.
The name of the gas is chlorine (clor-een).

Chlorine is an element. It smells like household bleach. It is poisonous. The symbol for chlorine is Cl.
It is a diatomic gas. Its molecule contains two atoms. It is written Cl_2.

So the misty gas is a compound of hydrogen and chlorine. It may contain other elements as well. To find out if it does, we let hydrogen and chlorine react.

*Experiment 11.5 Hydrogen and chlorine

1. Make hydrogen from zinc and dilute sulphuric acid. Fill a small gas jar. Fill a second jar with chlorine.

Put safety screens round the jars. Invert the chlorine jar over the hydrogen jar. Remove the covers from between them. Have a lighted splint ready. After a few seconds take away the top jar.

Put the lighted splint into the open bottom jar. Cover the top jar. Open it after several minutes.

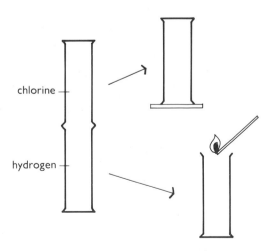

2. Use hydrogen from a cylinder. Pass it through the tube. Surround the experiment with screens. Light the gas. Lower it into a gas jar of chlorine. Hold a white paper behind the gas jar.

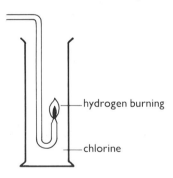

Does the lighted splint cause a change?
Do the gases hydrogen and chlorine react without it?
Is there any sign of the misty gas forming?
Does it appear when the top jar is opened?
Is chlorine used up as hydrogen burns in it?
Is one product the misty gas?

Does this experiment tell us what elements the misty gas contains?

If hydrogen and chlorine are mixed, they react. A lighted splint makes the mixture explode. So does sunlight. Hydrogen burns in chlorine. Chlorine burns in hydrogen. In all these reactions the misty gas is formed. It must contain hydrogen and chlorine only. It is hydrogen chloride.

The gas contains only two elements. Its name must end in -ide. It is hydrogen chloride.

We made it from concentrated sulphuric acid and salt. $NaCl\,(s) + H_2SO_4\,(l) \rightarrow NaHSO_4\,(s) + HCl\,(g)$

It is made in industry. Hydrogen is burnt in chlorine. $H_2\,(g) + Cl_2\,(g) \rightarrow 2HCl\,(g)$

Both gases are made by electrolysis.
Hydrochloric acid is a solution of HCl gas. So it can never be 100% acid. It must always contain water.

*Experiment 11.6   More about hydrogen chloride

Put small pieces of rock salt in the flask. Add concentrated sulphuric acid. Collect the gas as shown. This method is called 'downward delivery'.

a) Stand a gas jar of the gas over a gas jar of ammonia. Slide the lids out. Let the gases mix and stand for some time. Look at the bottom gas jar.

b) Fill a flask with the gas. Fit a stopper with glass tubing, narrowed at one end. Join rubber tubing to the other end. Close this with a clip.

Dip the narrow end of the tube in water. Put it into the flask. At the same time stopper the flask. Colour some water with Universal Indicator. Put the rubber tubing in the water. Open the clip.

What do you see when the gases mix?
What is left in the bottom gas jar after standing?
Water rushes into the flask. Why?

Ammonia and hydrogen chloride give dense white smoke. This settles as white crystals. They are ammonium chloride crystals. We recognize ammonia by this test.

We opened the clip. Water rushed up the tube. It entered the flask like a fountain. It must be filling 'empty' space. So most of the gas has gone. It must have dissolved in the water on the tube. A little water dissolves a lot of gas. Hydrogen chloride is very soluble. This is the 'fountain' test. About 800 cm³ of gas dissolve in 1 cm³ of water. The solution is concentrated hydrochloric acid.

This explains why the gas is misty in air. It dissolves in water vapour in the air. The mist is tiny drops of acid.

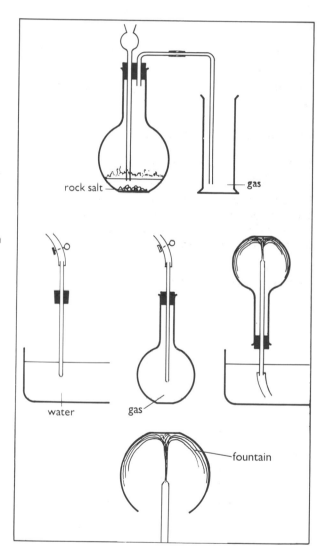

rock salt

gas

water   gas

fountain

Hydrochloric acid

Hydrogen chloride in water forms an acid. How does this happen. Is it an acid in other liquids?

*Experiment 11.7 Hydrogen chloride in another liquid

Put methylbenzene in a small flask. Add anhydrous calcium chloride. Stopper the flask. Let it stand overnight. Dry three test tubes and a litmus paper in an oven. Put the flask in a fume cupboard. TAKE CARE!

Pass dry hydrogen chloride into the methylbenzene. Let the gas bubble through for five minutes. Put some of this solution into each test tube. Add separately
a) the dry blue litmus paper,
b) a small lump of marble,
c) magnesium ribbon.

> Hydrogen chloride dissolves in methylbenzene. Is the solution an acid?
> Does it affect litmus, marble or magnesium?

Take the temperature of some water. Pour a little into all three test tubes. Stopper the tubes. Shake them. Take the temperature of the water in the litmus tube. Test gases formed in the other tubes.

> The water forms a bottom layer in each tube. Does it become hotter or colder?
> Does it react with litmus, marble or magnesium?

The methylbenzene solution is not acid. It does not affect dry litmus. It does not react with marble or magnesium. The water dissolves some of the hydrogen chloride. This This solution is acid. It affects litmus. It reacts with marble and magnesium. It is hydrochloric acid.

> water + hydrogen chloride → acid solution + heat
> water + pure sulphuric acid → acid solution + heat

Heat suggests chemical change. Perhaps both substances *react* with water.

*Experiment 11.8 Electrolysis of dilute hydrochloric acid

Repeat Experiment 10.2. Use dilute hydrochloric acid in the voltameter. First colour it with litmus. Use platinum electrodes. Collect the gases formed there.

> Check how fast the bubbles appear at each electrode. Compare them. Compare them with the volumes of gas collected. What gases are they likely to be?

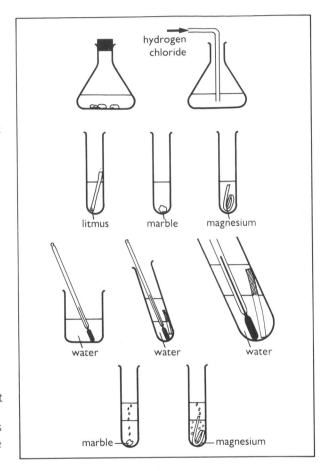

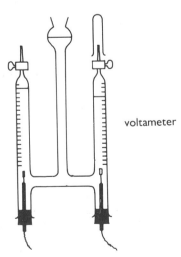

voltameter

What ions give hydrogen at the cathode?
What happens to the litmus round the anode?
What gas is being set free there?
What happens to the bubbles of this gas?

Hydrogen at the cathode must come from H^+ ions.
At the anode litmus is bleached. The gas must be
chlorine.
It comes off just as fast as hydrogen does.
Some of it is used up in bleaching. Some dissolves.
Very little of it reaches the tap.

The solution must contain H^+ and Cl^- ions.

at the cathode (negative)
$$H^+ + e \rightarrow H$$
$$2H \rightarrow H_2(g)$$

at the anode (positive)
$$Cl^- \rightarrow Cl + e$$
$$2Cl \rightarrow Cl_2(g)$$

Experiment 11.9 Other properties of the dilute acid

Fill test tubes one third full of dilute hydrochloric acid.
Drop into each tube one of the following
a) the metals magnesium, zinc and copper,
b) small amounts of magnesium oxide, zinc oxide and
    copper(II) oxide,
c) copper(II) carbonate, calcium carbonate.

When reaction stops, evaporate the solutions from c).
Heat the calcium chloride strongly. Drive off all the
water. Leave the solid standing in air.

Write a word equation for each reaction. Using the
formula HCl work out formula equations.
What happens to the solid in c)?

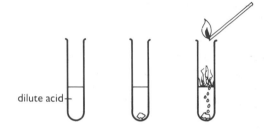

dilute acid

Experiment 11.10 To make salt crystals

Fill a burette with dilute hydrochloric acid. Fill the part
below the tap. Support it in a stand. Take the reading of
the acid level.

Put sodium hydroxide solution into a beaker. Pick up 25
cm^3 of it with a pipette. Use a pipette bulb. Run the
alkali into a conical flask. Add two drops of methyl
orange. Run acid into the conical flask.

Swirl the mixture. When the methyl orange shows signs
of changing, add acid in drops. Stop when the colour
changes to pink. Read the acid level.

Wash out the flask. Put in another 25 cm^3 of alkali. Run
in the same amount of acid as before. Evaporate the
solution in a dish. Stop when a little liquid is left.

Look at the crystals. What substance are they?

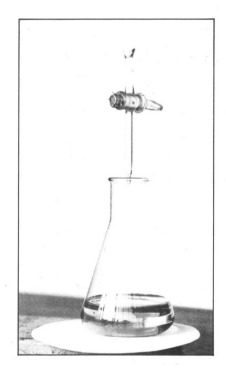

Chlorides

In water, hydrogen chloride turns into ions. It ionizes. The ions are H^+ and Cl^-. It is hydrogen ions which make a solution acid. Hydrogen chloride also dissolves in other liquids. It does not ionize in them. No acid forms.

$$HCl \rightarrow H^+ + Cl^-$$
hydrogen ' chloride
ion ion

The acid reacts with metals e.g. magnesium. The metal goes into solution as ions. Hydrogen appears as a gas.

$$Mg \rightarrow Mg^{2+} + 2e$$
$$2H^+ + 2e \rightarrow H_2$$

Mg^{2+} and Cl^- ions are left in solution. Evaporating it gives magnesium chloride crystals. This is how it happens. We can sum it up by writing

$$Mg(s) + 2HCl(aq) \rightarrow MgCl_2(aq) + H_2(g)$$

$$Mg + \begin{array}{c} 2H^+ \\ 2Cl^- \\ acid \end{array} \rightarrow \boxed{\begin{array}{c} Mg^{2+} \\ 2Cl^- \\ salt \end{array}} + H_2$$

Sodium hydroxide solution has Na^+ and OH^- ions.
Hydrochloric acid has H^+ and Cl^- ions.
They react. Water is formed. Na^+ and Cl^- ions are left in solution.
Evaporation gives sodium chloride crystals.

$$Na^+ + \boxed{\begin{array}{c} OH^- \\ H^+ \end{array}} + Cl^-$$
$$\downarrow$$
$$H_2O$$

It is useful to know that:
Most chlorides dissolve in water. Some do not. All nitrates and all sodium salts dissolve in water.

Experiment 11.11 An insoluble chloride

Take solutions of sodium chloride, zinc chloride and sodium carbonate. Add drops of silver nitrate solution. Pour in dilute nitric acid.

All three give a white precipitate (ppt).
Remember that all nitrates dissolve in water.
So do all sodium and potassium salts.

sodium chloride + silver nitrate $\rightarrow$? + ?

> In each case, which substance is the ppt?
> Does it dissolve in nitric acid?

sodium carbonate + silver nitrate $\rightarrow$? + ?

All chlorides give a white ppt of silver chloride.
The ppt does not dissolve in nitric acid.
Other salts give white ppts too.
They *do* dissolve in the acid.

Chlorides have Cl^- ions.
Silver nitrate solution has Ag^+ ions. They react.
We can use this test to recognize a chloride.

$$Cl^-(aq) + Ag^+(aq) \rightarrow AgCl(s)$$

The uses of chlorides

Sodium chloride is common salt. We use it to preserve and flavour food. Many other substances are made from it. Potassium chloride is a fertilizer. Ammonium chloride is used in dry cells. Anhydrous calcium chloride dries substances.

Oxidation of hydrochloric acid

Copper(II) oxide reacts with hydrochloric acid.
$$CuO + 2HCl \rightarrow CuCl_2 + H_2O$$
Copper chloride forms.
Manganese(IV) oxide is MnO_2. It has extra oxygen.
$$MnO_2 = MnO + O$$

$$MnO + 2HCl \rightarrow MnCl_2 + H_2O$$
$$O + 2HCl \qquad H_2O + Cl_2$$

$$MnO_2 + 4HCl \rightarrow MnCl_2 + 2H_2O + Cl_2$$

Hydrochloric acid is oxidized. The oxygen added takes away hydrogen. It gives water and chlorine.

Questions

1. A white sodium salt has cubic crystals. With concentrated sulphuric acid it gives a fuming, misty gas. This turns a wet litmus paper red. When mixed with ammonia it gives white smoke.

   Name the sodium salt, the misty gas and the white smoke. Why does a wet litmus turn red?
   Why is the gas misty in moist air but not in dry?

2. Some white crystals are thought to be common salt. They colour a bunsen flame yellow-orange. What does this mean?
   Two different tests will show if the substance is a chloride. Say exactly how you would carry out one of these tests.

3. What is neutralization?
   Common salt can be made by this method.
   Name the substances you would use.
   Describe briefly how you would use them.

Summary

All chlorides react with concentrated sulphuric acid. A misty gas is formed. Passing this gas over heated iron gives hydrogen. This proves that the misty gas is a hydrogen compound.

It is very soluble in water. The solution is an acid. All acids contain hydrogen. The misty gas must contain another element. Oxidizing the concentrated solution of it gives a green gas.

This is the element chlorine, symbol Cl. It is a diatomic gas. Its molecule contains two atoms. Hydrogen and chlorine react vigorously. The misty gas is formed. It must be a compound of hydrogen and chlorine only. It is hydrogen chloride, HCl.

Hydrogen chloride mixed with ammonia gives white smoke. It settles as solid crystals. They are the salt ammonium chloride.

Hydrochloric acid is a solution of hydrogen chloride in water. Other liquids dissolve hydrogen chloride. These solutions are not acids. They do not affect litmus, magnesium or marble.

Hydrochloric acid conducts. It can be electrolyzed. Hydrogen forms at the cathode. This means that the acid contains hydrogen ions, H^+.
An equal volume of chlorine forms at the anode. The acid must contain Cl^- (chloride) ions. HCl is still a useful formula for hydrochloric acid.

Like all acids hydrochloric acid forms salts. They are called chlorides. They are made by the usual methods. Acid + metal, metal oxide, metal hydroxide or metal carbonate. Most chlorides dissolve in water.

Chloride solutions give a white ppt with silver nitrate. This does not dissolve in nitric acid. It is used as a test for chlorides in solution.

$$X \text{ chloride} + \text{silver nitrate} \rightarrow X \text{ nitrate} + \text{silver chloride}$$

Chlorine

To make chlorine we oxidize hydrochloric acid. Concentrated acid is used. Manganese(IV) oxide oxidizes it.

*Experiment 11.12 The properties of chlorine

Use the apparatus in a fume cupboard. Put some manganese(IV) oxide in the flask. Add concentrated hydrochloric acid. Heat if necessary. Bubble the gas through water. This removes hydrogen chloride.

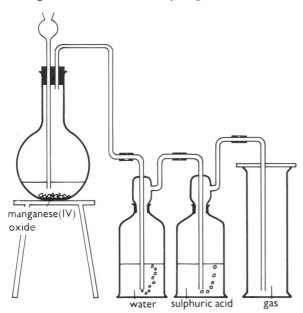

manganese(IV) oxide

water   sulphuric acid   gas

Pass the gas through concentrated sulphuric acid. This dries it. Collect the gas in gas jars as shown.

Lower into gas jars of chlorine
a) burning magnesium ribbon held in tongs;
b) sodium gently heated in a combustion spoon;
c) burning red phosphorus on a combustion spoon;
d) a burning candle on a combustion spoon.

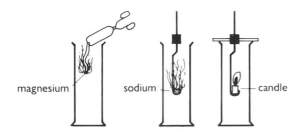

magnesium   sodium   candle

e) Stand two gas jars of chlorine in front of white paper. Add to one a few cm^3 of sodium hydroxide solution. Shake the mixture. Add a litmus paper.

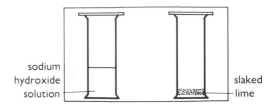

sodium hydroxide solution   slaked lime

To the second jar add newly slaked lime. Shake it in the jar. Remove the powder. Shake a little of it in water. Dip a litmus paper in the solution. Let two thirds of it get wet. Dip the other end in acid to the same depth.

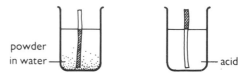

powder in water   acid

Which substances burn in chlorine?
What solid is formed when sodium burns?
Name two substances formed by the candle flame.
Does chlorine disappear when alkali is added?
Do the two react to form new substances?
What happens to the litmus papers?

Manufacturing chlorine.

Burning in chlorine

Chlorine reacts with metals. Magnesium burns in it. The solid product is magnesium chloride.
Sodium bursts into flame in chlorine. Cubic crystals of sodium chloride are formed.

Chlorine reacts with non-metals too. Phosphorus burns in it. Phosphorus chlorides are formed.

Candle wax is a hydrocarbon. It contains hydrogen and carbon only. It burns in chlorine. The flame is red and sooty. Soot is carbon. The other product is hydrogen chloride. We can write wax as $C_{16}H_{34}$.

$$Mg + Cl_2 \rightarrow MgCl_2$$

$$2Na + Cl_2 \rightarrow 2NaCl$$

$$2P + 3Cl_2 \rightarrow 2PCl_3$$

$$C_{16}H_{34} + 17Cl_2 \rightarrow 16C + 34HCl$$
candle   chlorine   soot hydrogen
wax   chloride

Chlorine and alkalis

Chlorine is a pale green gas. White paper makes it easier to see. It shows that alkalis remove the gas. They must be reacting with it. Remember that chlorine reacts with water. The solution is acid. It bleaches litmus paper white.

Chlorine and water form two acids. One bleaches. It is hypochlorous acid, HClO. Its salts are the hypochlorites. The acid and its salts bleach.

If water + chlorine → two acids
then alkali + chlorine → two salts

Sodium hydroxide and chlorine react. The solution is the household bleach we buy.

Calcium hydroxide and chlorine give 'bleaching powder'. Shaken with water it turns litmus blue. Acid makes the other end of the litmus paper red. Acid and bleaching powder reach the middle. They bleach it. We get a red, white and blue strip.

Chlorine belongs to a family. All the elements in it are called halogens. The word means 'salt former'. They are fluorine F, chlorine Cl, bromine Br, iodine I, astatine At. Their properties are very much alike. We know about chlorine. We shall test bromine and iodine.

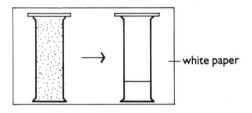

white paper

$$Cl_2 + H_2O \rightarrow HClO + HCl$$

$$HClO + coloured \rightarrow HCl + colourless$$
           dye   substance

$$Cl_2 + H_2O \rightarrow HClO + HCl$$
$$Cl_2 + NaOH \rightarrow NaClO + NaCl + H_2O$$

'hypochlorite' bleach

blue

blue   white   red

The halogen family

Fluorine and chlorine are pale green gases. Bromine is a brown liquid. Iodine is a shiny grey solid. Astatine is a solid. It is rare and radioactive.

*Experiment 11.13 Properties of bromine and iodine

a) **Put one drop of bromine in a test tube. Stopper the tube. Let it stand. Add a little water. Shake the mixture in the tube. Add a blue litmus paper. Add 1 cm³ of trichloromethane. Shake the tube.**

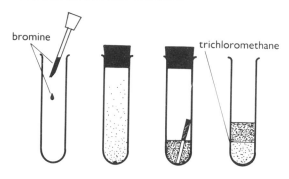

What happens to the drop of liquid bromine?
Does bromine dissolve in water and affect litmus?
What happens to the layer of trichloromethane?

b) **Put two or three iodine crystals in a test tube. Gently heat them. Let the tube cool. Add water. Stopper the tube and shake it. Add a blue litmus paper. Add a crystal of potassium iodide. Shake again. Shake the mixture with 1 cm³ of trichloromethane.**

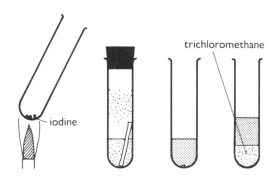

What happens when iodine is heated?
Does it dissolve in water?
What change does the potassium iodide make?
What happens to the layer of trichloromethane?

We know that chlorine dissolves slightly in water. The pale green solution bleaches litmus white.

Bromine turns into vapour. It also dissolves a little in water. The brown solution bleaches litmus slowly. Trichloromethane is more dense than water. It falls to the bottom. It becomes brown. It dissolves bromine.

Heated iodine turns into vapour. No liquid is formed. The iodine does not melt. We say that it **sublimes**. Iodine dissolves very little in water. The solution is very pale brown. It has little effect on litmus. Potassium iodide makes it dissolve much more. The solution is deep brown. It also dissolves in the trichloromethane layer. This solution is violet in colour.

*Experiment 11.14 Halogens and metals

a) **Pass chlorine over heated iron as shown.**

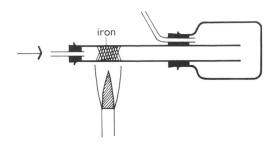

b) **Put two drops of bromine in a test tube. Put iron wool half way down the tube. Heat just below the iron. Do it again. Use iodine crystals instead of bromine.**

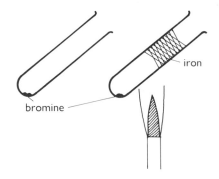

Do the three halogens react with iron? Put them in order of how well they react.

All halogens react with metals. Chlorine is the most vigorous of the three. Iodine is least active. Chlorides, bromides and iodides are formed. They are all salts.

Experiment 11.15 Chlorides, bromides and iodides

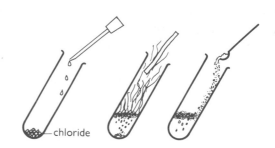

chloride

*a) Put sodium chloride crystals in a test tube. Add drops of concentrated sulphuric acid. Hold a wet litmus paper in the gas. Add a pinch of manganese(IV) oxide. Test the gas with litmus paper again. Add concentrated sulphuric acid to a bromide and an iodide. Test the gases with wet litmus paper.

> All three give misty gases. Name them.
> What else appeared in the bromide tube?
> What other product forms in the iodide tube?
> What effect does the manganese(IV) oxide have?

A chloride gives misty hydrogen chloride. Bromide and iodide give misty gases. They must be hydrogen bromide and hydrogen iodide. Manganese(IV) oxide oxidizes hydrogen chloride. Chlorine is formed.

Brown bromine forms in the bromide tube. Grey iodine forms in the iodide tube. The misty gases have been oxidized. The sulphuric acid does this.

b) Use solutions of chloride, bromide and iodide in water. Do the tests below on each one.

1. Add drops of silver nitrate solution. Add a little dilute nitric acid. Pour in aqueous ammonia.

> What colour is each ppt?
> Are they affected by the acid?
> Does the aqueous ammonia cause any change?

2. Add chlorine water to each solution. Put in 1 cm^3 of trichloromethane. Shake each tube gently.

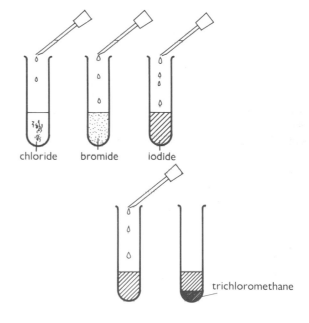

chloride   bromide   iodide

trichloromethane

> What effect does chlorine have in each case?
> Compare the trichloromethane in each tube.

All three give ppts with silver nitrate. Nitric acid does not affect them. It dissolves other silver ppts.
The silver chloride ppt is white. Ammonia dissolves it. Silver bromide is cream coloured. Ammonia nearly dissolves it. Pale yellow silver iodide is unchanged.

Chlorine is the most active. It displaces bromine and iodine. Both solutions are brown. They look alike. Trichloromethane tells us which is which.
Bromine makes it brown.
Iodine makes it violet.
Chlorine does not react with chlorides. Nothing new is formed. Trichloromethane is not changed.

Suppose we have crystals of a salt. It may be a chloride, bromide or iodide. These tests show us which it is.

175

Happy families?

We have tested chlorine, bromine and iodine. We cannot test fluorine or astatine. The halogens are very much alike. So are their compounds. They react in the same kind of way. In our tests chlorine was the most active.

Bromine comes next. Iodine is least active. This is why chlorine reacts with bromides. Chlorine pushes out the less active bromine. It also displaces iodine. Both solutions become brown. We used trichloromethane to tell us which was which. It dissolves bromine. The solution is brown. It dissolves iodine. The solution is violet.

The compounds of the halogens are alike. A summary shows this.

Reaction with	Chloride	Bromide	Iodide
1. concentrated sulphuric acid	misty hydrogen chloride forms	misty hydrogen bromide + bromine	misty hydrogen iodide + iodine
2. chlorine water add trichloromethane	no reaction no change	brown solution of bromine brown solution in this liquid	brown solution of iodine violet solution in this liquid
3. silver nitrate solution	white ppt unchanged by acid dissolves in ammonia	cream ppt unchanged by acid nearly dissolves in ammonia	pale yellow ppt unchanged by nitric acid or ammonia

All three tests can be used to test for chloride, bromide or iodide. All three give similar results. They differ enough to say which compound we started with.

Every element belongs to a family. Lithium, sodium and potassium do. They are called the alkali metals. As in all families their properties are alike. Potassium is the most active. Lithium is least active.

Look at their relative atomic masses or mass numbers.
   Li = 7, Na = 23, K = 39
Can you see a connection?

Summary

To make chlorine we use concentrated hydrochloric acid. The change is oxidation. Manganese(IV) oxide can oxidize the acid. Chlorine dissolves in water.

Chlorine reacts with metals and non-metals. The compounds formed are called chlorides. Sodium and magnesium both burn in chlorine. The equations are

$2Na(s) + Cl_2(g) \rightarrow 2NaCl(s)$

$Mg(s) + Cl_2(g) \rightarrow MgCl_2(s)$

Other metals heated in chlorine react.

Phosphorus is a non-metal which burns in chlorine. Candle wax is a hydrocarbon. It is a compound of hydrogen and carbon. A candle burns in chlorine. It has a smoky flame because soot is formed. Soot is carbon. Hydrogen chloride is also formed.

Chlorine in water forms two acids.

$Cl_2(g) + H_2O(l) \rightarrow HCl(aq) + HClO(aq)$

So with alkali chlorine forms two salts.

$Cl_2 + 2NaOH \rightarrow NaCl + Na\ ClO + H_2O$

Both solutions bleach. HClO is hypochlorous acid. NaClO is sodium hypochlorite. They do the bleaching. They oxidize the coloured dye.

$NaClO + dye \rightarrow NaCl + colourless\ substance$

Slaked lime is an alkali, calcium hydroxide. It reacts with chlorine. It forms bleaching powder.

Chlorine is a member of the halogen family. The others are fluorine, bromine, iodine and astatine. They have similar properties. Chlorine is more active than bromine. Bromine is more active than iodine. Chlorine reacts with bromides and iodides.

It displaces bromine and iodine. They appear as brown solutions. Trichloromethane dissolves both. Bromine gives a brown solution, iodine a violet one.

Chlorides, bromides and iodides are alike. All three give a misty gas with concentrated sulphuric acid. The gases are hydrogen chloride, hydrogen bromide and hydrogen iodide. The last two are oxidized by the acid. So bromine and iodine are formed as well.

All three, in solution, give a ppt with silver ions. The ppts do not dissolve in nitric acid. The chloride ppt dissolves in ammonia. The bromide ppt nearly dissolves. The iodide ppt is not soluble.

Every element belongs to a family. The members of the family are alike. Their properties are similar. Some members react more vigorously than others.

Questions

1. Fill in the blanks.

   Chlorine can be made from concentrated _____ acid. This is an example of an _____ reaction. Manganese(IV) oxide provides the _____. Chlorine is _____ in colour. Sodium _____ in chlorine. This gives white crystals of _____ _____. Some metals react if _____ in chlorine gas.

   A non-metal which burns in chlorine is _____. Hydrocarbons such as candle wax contain _____ and _____. They react with chlorine. The products are hydrogen _____ and soot. Soot is the element _____.

   Chlorine dissolves in water. The solution turns litmus _____. It then _____ it white. The first change shows that the solution is _____.
   $Cl_2 + H_2O \rightarrow HCl +$ _____. The second acid does the _____. It is hypochlorous _____ or chloric(I) acid.
   $HClO + dye \rightarrow HCl +$ _____ _____.

2. A salt A consists of white cubic crystals. Concentrated sulphuric acid is added to A. The gas B is formed. It is misty in moist air. It turns litmus red. Mixed with ammonia it gives white smoke. This settles as crystals of substance C. Manganese(IV) oxide is added to the sulphuric acid and salt mixture. A green gas D is formed. It will bleach a moist litmus paper.
   Name the salt A and the gases B and D.
   What is the white smoke C?
   Why does D bleach litmus paper?
   How would you make a sample of household bleach from gas D?

Chemistry and disease

A disease makes us become ill. Our bodies do not 'work' normally. The effects are called symptoms. For example, body temperature may rise above normal. There are four main causes of disease:
1. The body is short of a substance it must have,
2. The body is invaded by tiny living creatures,
3. A part of the body is not normal. It may have been damaged. It may always have been like it.

Missing substances

A missing substance causes a **deficiency disease.** Scurvy is an example. It was once very common. It causes bleeding gums. It can start internal bleeding. Many people died of it in the past.

In the days of sailing ships, voyages were long. The sailors suffered from scurvy. So did people in poor city areas. They were cured by eating fresh vegetables or fruit. Sailors far from land had none. Poor city people could not afford them.

Lime juice was first carried by British ships. It prevented scurvy. British sailors are still called 'limeys'. Tests proved that fruits contain an essential substance. So do fresh vegetables. It is ascorbic acid, vitamin C. It can be manufactured. We can buy vitamin C tablets. A good diet gives other essential substances. They prevent other deficiency diseases. Poor countries still have them.

The body is a chemical factory. It makes very many substances. One is insulin, made in the pancreas. Without it the body cannot use up sugar. So sugar builds up in the blood. It acts as a poison which can be fatal. The disease is diabetes. A faulty pancreas causes it. Man-made insulin cures it.

Tiny invading creatures

Bacteria and viruses are common ones. Bacteria can be seen under a microscope. Viruses cannot. Both are **micro-organisms.** There are several other kinds. They enter the body. They live and multiply there. Some micro-organisms are helpful. It was proved over 100 years ago that many others cause

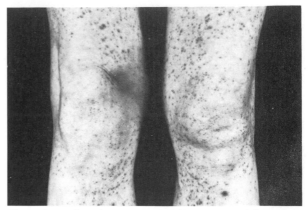
The scabs on these legs were caused by scurvy, a deficiency disease.

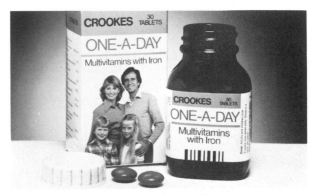
Vitamin tablets help prevent deficiency diseases.

disease. The body has defences against them. It produces antibodies in the blood. They attack and destroy micro-organisms. So do white blood cells.

Some compounds can destroy micro-organisms. Some were in use in very early times. They came from plants. For example, Indians in Peru suffered from malaria. They chewed the bark of the cinchona tree. This could cure the disease and prevent it. The bark was found to contain a substance called quinine. It could be manufactured. It took the place of the bark.

Compounds which kill micro-organisms are called **antibiotics.** (It means 'against living things'). Today most of them are man-made. Some come from living things. The most famous one is penicillin.

Spraying crops with a pesticide.

Some villages rely on one water supply. It is disastrous if this water becomes infected.

Growing a mould called penicillium produces penicillin. It destroys many types of micro-organisms. So many diseases are no longer common. Diptheria is one.

A part of the body not normal

Compounds may help to put it right. It may need surgery. If so, the person is 'put to sleep'. Substances which do this are called **anaesthetics.**

Yes, there is a fourth cause!

The body can become ill. Mental illness can happen too. Body and mind work together. Each affects the other. Mental illness can have physical effects.

Other ways of controlling disease

If organisms enter the body, it is 'infected'. Some enter through wounds. We can destroy those around a wound. We use substances called **antiseptics.** We destroy them elsewhere with **disinfectants.** Both reduce the risk of infection. They reduce disease.

Disease organisms get into food and drink. They come from dirty, infected hands or containers or kitchens. Dirt means disease. Wash well with soap. Coughs put water drops into the air. They contain micro-organisms. They fall on food. Other people breathe them. Coughs and sneezes spread diseases.

Insects hand on disease. Mosquitoes carry malaria organisms. A bite puts them into the bloodstream. The person bitten gets the disease. Lice and fleas are carriers. They live and travel on animals.

To reduce disease, kill the carriers. Substances like DDT kill insects. They are called **pesticides.** They have drawbacks. They can kill useful insects. They harm other wild life, such as birds. Pests build up resistance to them. They become harmless.

Social medicine

Prevention is better than cure. Micro-organisms thrive and are passed on best
a) in overcrowded living conditions,
b) by infected food and drink,
c) by poor sewage disposal,
d) by poor hygiene and health habits.

Poor diet causes deficiency diseases. It can cause others such as heart disease. It makes resistance to disease weaker. People get better less easily.

Britain has changed in the last 100 years: Living conditions have become better. Almost everyone has a pure water supply (page 102). Sewage is dealt with. Refuse is collected (page 103). People are taught about health and diet. Baths and W.C.s make it easier to be clean. The National Health Service has grown. There are new ways of healing mental illness.

Prevention and cure work side by side.

How do elements combine?

Chlorine is a green gas. Sodium is a metal. They are both elements. Like all elements they contain atoms. Sodium burns in chlorine. Their atoms combine.

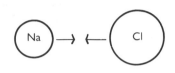

To do this the atoms must meet. They end up as white salt crystals. How do they combine? Check again on what you know about atoms.

Every atom has a **nucleus**. It is very small compared with the atom. It contains protons and neutrons. Protons give it a positive charge. Neutrons have no charge. Both have the mass of a hydrogen atom.

nucleus = protons + neutrons

An atom contains **electrons.** They are equal in number to its protons. They have a negative charge. They are in groups called **shells.** These surround the nucleus. You need to know more about shells.

electrons in shells

The first shell is the one nearest to the nucleus. It can hold at most only two electrons.

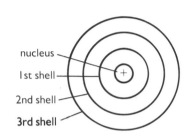

The second shell is further from the nucleus. It can hold up to eight electrons. It is then full. The third shell is full when it contains eighteen.

Apply these ideas to the first twenty elements.

Hydrogen   Atomic Number 1. Mass Number 1.
Its atom has one proton in the nucleus.
It has one electron in the first shell.

Helium   Atomic number 2. Its nucleus has two protons.
Mass Number 4. The nucleus has two neutrons.
It has two protons so it will have two electrons as well.
They are both in the first shell. This shell is then full.

Lithium   Atomic number 3. Mass Number 7.
Its atom contains three electrons. Two of these fill the first shell. The third will be in the second shell. This arrangement of electrons is called the electron structure. It is written 2.1.

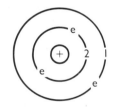

The table opposite shows twenty elements. Only the *electron* structure of each is worked out. Can you see why the *nucleus* is less important?

Pass the salt, Willie.

When sodium, soft and shiny metal,
Meets chlorine, green and poisonous gas,
The two combine and crystals settle.
So now we have some salt to pass!

Element	Atomic Number	Number of electrons in each shell				
		1st	2nd	3rd	4th	5th
Hydrogen	1	1				
Helium	2	2 ← (shell full)				
Lithium	3	2	1			
Beryllium	4	2	2			
Boron	5	2	3			
Carbon	6	2	4			
Nitrogen	7	2	5			
Oxygen	8	2	6			
Fluorine	9	2	7			
Neon	10	2	8 ← (shell full)			
Sodium	11	2	8	1		
Magnesium	12	2	8	2		
Aluminium	13	2	8	3		
Silicon	14	2	8	4		
Phosphorus	15	2	8	5		
Sulphur	16	2	8	6		
Chlorine	17	2	8	7		
Argon	18	2	8	8 ← (part shell full)		
Potassium	19	2	8	8	1	
Calcium	20	2	8	8	2	

hydrogen

helium

lithium

magnesium

Atoms of sodium and chlorine combine. To do so they must meet. The shells surround the nucleus. **So the shells meet first.** The nucleus is less important. The shells must control combination.

Use the table to find answers to these questions.

1. Which two elements have full outside shells?
2. Which other element has eight electrons in its outside shell?
3. The list has three alkali metals. Which are they?
4. Check their outside shells. How many electrons in each?
5. Which is the outside shell of chlorine, 1st, 2nd or 3rd?
6. How many electrons does this outside shell contain?
7. What other element has the same number in its outside shell?
8. Is this element a halogen?

The first shell is filled by two electrons. So helium has a full outside shell. The second shell is full when it has eight electrons. Neon has a full outside shell. Both are noble gases. So is argon. Its structure is 2.8.8.

He 2
Ne 2.8
Ar 2.8.8

The alkali metals are lithium, sodium and potassium. Their atoms each have an outside shell with one electron. Chlorine has the electron structure 2.8.7. Its outside shell has seven electrons. So does the fluorine atom, 2.7. Both are halogens.

Li 2.1
Na 2.8.1
K 2.8.8.1

The ionic bond

All atoms of noble gases have either
  a) a full outside shell or
  b) an outside shell of eight electrons.
All other atoms tend towards this. By combining they reach the structure of a noble gas atom.

The electron structure of a sodium atom is 2.8.1.
The nearest noble gas is neon, structure 2.8.
To reach this, the sodium atom loses one electron.
It now has only 10 electrons. Their charge is 10−.
The atom still has 11 protons. Their charge is 11+.
It has one more positive than negative charge.
It has become a sodium ion, Na^+, structure 2.8.

A chlorine atoms has the electron structure 2.8.7.
The nearest noble gas is argon, structure 2.8.8.
A chlorine atom therefore gains one electron.
Now it has 18 electrons. Their charge is 18−.
It still has 17 protons. They have a charge of 17+.
The atom now has an extra negative charge.
It has become a chlorine ion, Cl^-, (2.8.8.).

Can you see how a reaction works?
A small piece of sodium burns in chlorine. One sodium atom loses an electron, e. It becomes Na^+.
A chlorine atom gains this electron. It forms Cl^-.

These two ions have opposite charges. They will attract each other. They join up as $Na^+.Cl^-$.

Another sodium atom loses an electron. It forms an Na^+ ion. Which of the ions already formed will attract it? Will the arrangement become $Na^+.Cl^-.Na^+$ or $Na^+.Na^+.Cl^-$?

Forming a second Na^+ gives an electron.
This turns another chlorine atom into an ion, Cl^-.
Which ions will attract it?

Burn 23 g of sodium. It contains L atoms. They lose L electrons. L ions of sodium are formed. L atoms of chlorine gain an electron each. L ions of chlorine form. Na^+ and Cl^- ions join up in all directions. 58.5 g of salt form. It will be in the form of cubic crystals.

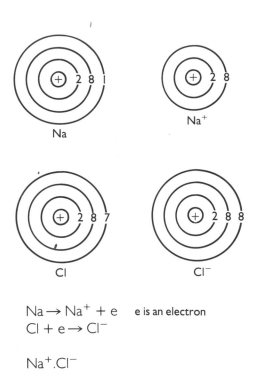

$$Na \rightarrow Na^+ + e \quad \text{e is an electron}$$
$$Cl + e \rightarrow Cl^-$$

$$Na^+.Cl^-$$

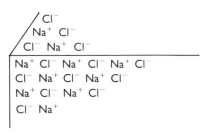

This arrangement of ions is called a crystal lattice.

Now you can see why electron structure is important. The most important part is the outside shell. The outside shell is what other atoms first bump into. It decides whether the atom gains or loses electrons. It also decides how many electrons it gains or loses.

Look again at the table on page 181. Make a list of the metals in it. Check the number of electrons in their outside shells. Is the number small or large?

Will all metal atoms lose electrons or gain them? Will they form positive or negative ions? Will metal ions be attracted to the cathode or anode?

List the non-metals in the table. Check their outside shells. Is the number of electrons small or large? Will they gain or lose electrons?

Non-metals gain electrons. Their ions carry a negative charge. They are attracted to the anode.

Hydrogen is unusual. Its outside shell has one electron. Like metals it can lose this. It will become a positive ion, H^+. The nearest noble gas is helium. A hydrogen atom can also gain one electron. It becomes the ion H^-.

Attraction between the ions is called the **ionic bond.** The compounds it forms are **ionic compounds.**

Here is another example. Copy it into your book. At the same time fill in the blanks.

Magnesium burns in chlorine. The compound formed is magnesium _____. Magnesium is a _____. The electron structure of magnesium is 2.8. _____. The nearest noble gas is neon, structure 2._____. So a magnesium atom will lose _____ electrons. It loses them from its outside _____. It now has _____ electrons and _____ protons. It has _____ more positive than negative charges. It has become a magnesium ion, Mg^{2+}.

A chlorine atom has the structure 2.8. _____. Its outside shell contains _____ electrons. The nearest noble gas is _____, structure 2.8.8. So a chlorine atom _____ one electron. It becomes the _____tively-charged ion, Cl^-. A magnesium atom loses two electrons. These turn _____ Cl atoms into ions. Magnesium chloride is Mg^{2+}. $2Cl^-$ or $MgCl_2$.

Ionic bonding explains why:
    One sodium atom reacts with one chlorine atom.
    A magnesium atom reacts with two chlorine atoms.
    Ionic compounds are solids. They from crystals.

Summary

Atoms contain electrons. They are in one or more shells. The outside shell is the most important. When atoms collide their outside shells meet. These shells lose or gain electrons. They reach noble gas structure. The atoms form ions.

Metals lose electrons. They form positive ions. Non-metals gain electrons. They have negative ions. Ions of opposite charge attract each other.

This holds them together in huge numbers. They form solid crystals. These have a definite shape. Attraction between ions is the ionic bond. Compounds made of ions are ionic compounds.

Answers

chloride	two
metal	2.8.7.
2.8.2.	seven
2.8.	argon
two	gains
shell	negatively
10 electrons and	
12 protons	

Why do elements combine?

We have seen how elements can combine. They do so by the transfer of electrons. One atom loses them. Another atom gains them. Both become ions.

The ions have opposite charges. They attract each other. Huge numbers join up. This is ionic bonding. The result is a solid crystal lattice.

This is **how** it happens. **Why** do atoms do this?

The chief reason is simple. Everything tends to lose energy. Energy is what is needed to do work. In Science, **work is done when a force moves through a distance**.

Take a simple example.

My case is on the floor. I need it on the table. The earth pulls it downwards. This pull is the weight of the case. To lift it, I must use a force bigger than its weight. This force moves through a distance. In lifting the case I have done some work.

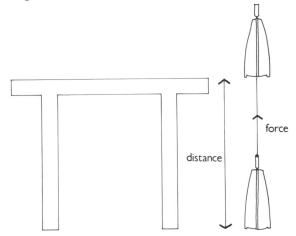

The same work can be done in many other ways. George falls asleep in Science lessons. Tie a rope to him. Pass the rope over pulleys. Tie the other end to my case. Let George fall asleep. Put a mattress behind him. Give him a gentle push.

He falls off his stool. This pulls the rope. It lifts the case to the table top. George has done work. So he must have used energy. What energy?

George had energy because of his position. It is called **potential energy.**

He lost it when he fell. It was used to do work. It raised my case to the table top.

On the floor he has no potential energy. He can do no work. My case has gained potential energy.

Heat is another form of energy. We can use it to boil water. The steam given off will drive an engine. The engine makes a wheel go round. The wheel has energy of movement. This is called **kinetic energy**. We can use it to lift the case.

I can put an explosive under my case. Setting it off will lift the case. It may also lift the table. It could blow both to pieces.

Explosives contain chemical compounds. Energy stored in compounds is called **chemical energy**.

No energy is ever lost. It is just turned into a different form. One kind of energy can be turned into any other kind. Chemical energy of an explosive is turned into

kinetic energy (it makes things move),
heat energy (from the reaction),
sound energy (the bang),
light energy.

All things tend to a lower energy level. Substances do. They give up energy when their atoms react. This appears mainly as heat.

Noble gases have no simple reactions. Their atoms must already be at a low energy level. They are like George on the floor. They cannot fall further.

This explains why other atoms lose or gain electrons. They become like noble gas atoms. They reach a lower energy level. The 'lost' energy appears as heat.

Burn magnesium in oxygen. The electron structure of magnesium is 2.8.2. The nearest noble gas is neon, 2.8. A magnesium atom loses 2 electrons. It gives up energy. It becomes a magnesium ion, Mg^{2+}.

Oxygen has electron structure 2.6. Again the nearest noble gas is neon, 2.8. An oxygen atom therefore gains 2 electrons. It gains those given by the magnesium atom. It becomes the ion O^{2-}. By becoming like a neon atom it too gives up energy.

The two ions have potential energy. They attract each other. In moving together they can do work. Can you imagine them lifting my case? In fact the energy of all three changes appears as heat.

A rocket turns chemical energy into kinetic energy.

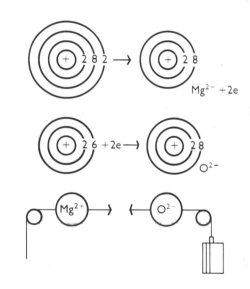

Burning petrol in a car engine gives the car energy to move.

Atoms turning into ions give heat. Ions joining up to form a crystal lattice give heat. This is why magnesium burns with a white hot flame. Other substances react with oxygen. Fuels burn. They give energy. It appears as heat. We can turn heat energy into other useful forms.

Petrol burns in motor engines. Gas and heat form in the cylinder. Gas molecules have kinetic energy. They hit the piston. It goes down. Its up-and-down movement rotates an axle. This turns the car wheels.

chemical energy $\rightarrow$ heat $\rightarrow$ kinetic energy
of petrol   of moving car

185

Another kind of bond

Atoms combine only if they meet. An atom in Bath cannot combine with an atom in Aberdeen. When they meet their outside shells meet first. Both shells contain electrons. In ionic bonding these are transferred from one atom to the other. Ions form.

Atoms can combine by a second method. They can share electrons. Their outside shells overlap. By sharing electrons each reaches noble gas structure. The need to share holds the atoms together. Take hydrogen as an example.

Two hydrogen atoms move towards each other. Their outside shells overlap. Each contains one electron. So each shell has a share of two electrons. Each atom has noble gas structure but only if they stay together.

The atoms are held together by a shared pair of electrons. Each atom provides one of the pair. The shared pair is called a **covalent bond.** Substances held together by covalent bonds are **covalent compounds.**

All electrons are exactly alike. In drawings, one electron is shown as x. The other is shown as o. This shows which atom each comes from.

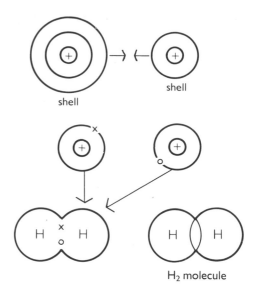

shell shell

H_2 molecule

The properties of covalent compounds

What properties does this bonding suggest?

It explains why hydrogen has diatomic molecules.

The two H atoms are identical. They will share the electron pair equally. So neither atom has an electrical charge. Nor has the H_2 molecule. So hydrogen molecules do not affect each other. They can move freely. This explains why hydrogen is a gas. Most liquids and gases are covalent. Those with large covalent molecules are solids.

Do you understand covalent bonding?
If not, this may help.
Grace keeps a rabbit. Its name is Electron x. It lives in a pen. The pen is called Shell G. Willie has a rabbit too. Its name is Electron o. It lives in a pen. This is called Shell W.

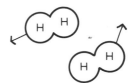

Grace and Willie would both like two rabbits. Each pen can hold two. Willie and Grace decide to share rabbits. They let the pens overlap. The rabbits occupy both shells. Each shell has a share of two rabbits, but only if they stay joined.

Try this example. Fill in the blanks. The drawings will help you.

The electron structure of chlorine is _____. The outer shell of its atom has _____ electrons. To be like argon, 2.8.8., it needs _____ electrons.

A hydrogen atom has _____ electron. It can share one _____ pair with a chlorine atom. It then has 2 electrons. It is like the noble gas _____.

The Cl atom now has an outer shell of _____. H and Cl are held together by an electron _____. A shared pair is a _____ bond. The formula of hydrogen chloride is therefore _____. Like many covalent compounds it is a _____.

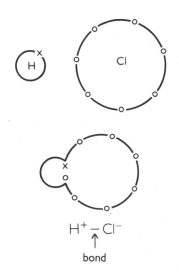

A covalent bond is shown by a short line. It stands for a pair of electrons. Each atom provides one.

A hydrogen molecule contains two atoms. They are held together by a shared pair. They share them equally.

An HCl molecule contains two different atoms. They do not share the electron pair equally. A chlorine atom attracts electrons. The electron pair is more with chlorine. The Cl atom has a slight negative charge. The hydrogen atom will have a positive charge. $H^+ - Cl^-$

Summary

Atoms can combine by sharing electrons. By doing this they reach noble gas structure. They have fallen to a lower energy level.
Each shared pair is called a covalent bond. Each atom provides one electron of the pair. A line represents the bond.

Substances with this bonding are covalent compounds. In some the atoms share the pair equally. Neither atom carries a charge.
They do not attract other molecules. They move freely. They are gases or liquids.
If they have large molecules they may be solids. Ionic bonding gives crystalline solids.

Grace likes rabbits more than Willie does. She attracts them. The rabbits are more often in Shell G. It has a bigger share of them than Shell W.

Atoms share electron pairs. They may not share equally. One atom may attract electrons. It will take a larger share. This gives it a negative charge. The other atom has a positive charge.

Covalent bonding explains why
a) some elements have diatomic molecules
b) some substances are gases or liquids.

Questions

1. Write down the electron structure of an oxygen atom. Use it to show that oxygen molecules are diatomic. Work out the formula of hydrogen oxide.

2. What is a covalent bond? Compare covalent with ionic bonding. Which compounds are more likely to be solids?

187

Ionic and covalent compounds

We have shown two ways in which atoms combine. One method is by the transfer of electrons. Loss and gain of electrons turns the atoms into ions. These collect in crystal lattices. Ionic compounds are solid.

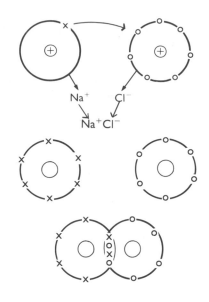

In the second method atoms share electrons. Each shared pair is a covalent bond. The bond holds the atoms together. Molecules are formed. They are able to move freely. Covalent compounds are liquids or gases. Only those with large molecules are solids.

Sodium chloride is the typical ionic compound. Candle wax is a solid covalent compound. We shall compare their simple properties.

Experiment 12.1 Ionic and covalent properties

Set up the apparatus shown. Switch on the current. Touch the carbon rods together.

a) Take a lump of solid wax. Press a carbon rod against one side of it. Press the other rod to the opposite side. Test a lump of salt in the same way.

*b) Put wax into a crucible. Heat it gently on a pipeclay triangle. Hold the carbon rods in the molten wax. Switch on. Bring the rods close together.

Gently heat salt in a crucible. Then heat it more strongly. If it melts, dip the rods in the liquid. Switch on the current. Watch the two carbon rods.

c) Put wax and salt in separate beakers. Add water and stir. Use the carbon rods to find out if either liquid is a conductor.

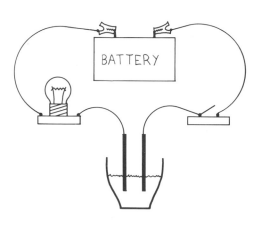

> You have seen some of these tests before.
> Is solid wax or solid salt a conductor?
> Which has the higher melting point, wax or salt?
> Is molten wax a conductor or an insulator?
> Does the bulb light up when the rods are in molten salt?
> Does wax dissolve in water?
> Does salt solution conduct a current?

When the rods touch the bulb lights up. A current flows. With wax between the rods the bulb does not light up. Solid wax is a non-conductor. So is solid salt.

Wax melts with gentle heating. Its melting point is low. Its molecules are easy to separate. They do not carry a current.

Salt is difficult to melt. Melting breaks up the crystal lattice. Na^+ and Cl^- ions have to be pulled apart. This needs a lot of energy. It comes from very strong heating. So the melting point is high. It is over 800°C.

Ions move freely in the liquid. Electrolysis can occur. Na^+ ions move to the cathode. Sodium metal forms there. Cl^- ions move to the anode. Chlorine gas is set free.

Salt dissolves in water. the Na^+ and Cl^- ions separate. They can move toward cathode and anode.
The solution conducts a current. Electrolysis can happen. The current consists of moving ions.

Wax does not dissolve in water. We can try other covalent compounds which do.

Two atoms as a general rule
Join up to form a molecule.
What keeps them linked together there
Is a shared electron pair.
For this covalent bond the sign
Is the two symbols joined by line (C – H).

Experiment 12.2 Solutions of covalent compounds

Use the apparatus of Experiment 7.1. Put ethanol (meths) In a beaker. Dip the carbon rods into it. Switch on. Take the temperature of the ethanol. Pour an equal volume of water into the beaker. Take the temperature again.

Take the temperature of some water. Dissolve hydrogen chloride in it. Take the temperature again. Find out if this solution is a conductor.

Is ethanol a conductor?
Does it dissolve completely in water?
Does the temperature change?
Is the solution able to conduct a current?
Does hydrogen chloride solution conduct?
Are new substances formed at the cathode and anode?

Ethanol is not a conductor. It dissolves completely in water. The solution is not a conductor either. Hydrogen chloride dissolves in water. The solution conducts. It is dilute hydrochloric acid. The temperature rises slightly. Perhaps a reaction is taking place.

Ionic compounds are solids. They have high melting points. The molten liquid conducts. So does the solution in water.
Covalent compounds are gases or liquids. Those with large molecules may be solids. They have low melting points. Some dissolve in water. Some of the solutions conduct. This means that some of them form ions in solution.

Acid, alkali and pH

Electrolysis of any acid solution gives hydrogen. It appears at the cathode. This is the negative electrode. So hydrogen ions must be positive, H^+.

But acids have covalent molecules. They dissolve in water. How do they put H^+ ions into solution? On page 59 we proved water to be hydrogen oxide. We gave its formula as H_2O. Is this correct? We can test it by experiment. Theory may confirm it.

*Experiment 12.3 The formula of water

Dry both copper(II) oxide and anhydrous calcium chloride in an oven. Measure the mass of a) the tube and copper oxide, b) the U-tube and chloride. Set up the apparatus as shown. Pass hydrogen through. Collect it at the other end. Test it with a lighted splint. When it burns quietly, light it at the jet.

Heat the oxide. Reduce it all to copper. Let it cool. Keep hydrogen passing through. Measure the new mass of a) the U-tube and chloride, b) the tube and copper.

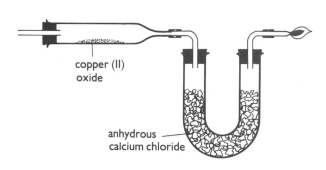

copper (II) oxide

anhydrous calcium chloride

loss in mass of oxide = 2.00 g
total mass of water   = 2.25 g
mass of hydrogen   = 0.25 g

Hydrogen takes oxygen from copper oxide. Water forms. The chloride absorbs it. Its gain in mass is water. The oxide loses mass. This loss is the oxygen.

2.00 g of oxygen combine with 0.25 g of hydrogen.
16 g of oxygen combine with 2 g of hydrogen.
  1 mole of oxygen atoms : 2 moles of hydrogen atoms
So the formula of water is H_2O. Theory confirms it.

The electron structure of oxygen is 2.6. For neon it is 2.8. So an oxygen atom gains two electrons.

A hydrogen atom provides one. Each oxygen atom needs two H atoms. They form two shared pairs. A water molecule is H_2O. It has 2 covalent bonds.

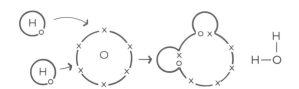

The O in the H-O attracts electrons. It has a bigger share of each shared pair. It is negatively charged. Dissolve hydrogen chloride in water. HCl and H_2O molecules meet. The O of H_2O attracts H away from HCl. The Cl keeps the shared pair. This makes it a Cl^- ion. The H, minus its electron, is an H^+ ion.

Sodium hydroxide is an alkali. It is a solid. It contains Na^+ and OH^- ions. Ions separate when the solid dissolves. Alkalis put OH^- or hydroxyl ions into solution. These ions cause alkaline properties.

$$Na^+ \ OH^- \ Na^+ \ OH^- \ Na^+$$
$$OH^- \ Na^+ \ OH^- \ Na^+ \ OH^-$$
$$Na^+ \ OH^- \ Na^+ \ OH^- \ Na^+$$

An indicator is a coloured substance. Some of them come from plants. Litmus is one of these. It is affected by acid or alkali. Hydrogen ions, H^+, make it red. Hydroxyl ions, OH^-, make it blue. The colour shows us that acid or alkali is there. It does not tell us how much.

Concentration (how much) is important. For example, hydrogen chloride is HCl. Its formula mass is 36.5. 36.5 grams of it is one mole. It contains L molecules. H = 1; Cl = 35.5; HCl = 36.5

Put this amount into water. Make the volume up to 1000 cm³.
This is a Molar, or M, solution. In it HCl molecules have turned
into ions. It ionizes well. Every molecule gives a hydrogen ion.
1000 cm³ of M hydrochloric acid contain L hydrogen ions.

HCl in water → H⁺ (aq) + Cl⁻ (aq)

Vinegar contains ethanoic or acetic acid. A Molar solution contains L
molecules. Only about 2% of them turn into ions. A Molar solution
has far fewer H⁺ ions. Hydrochloric acid has 50 times as many. It is a
strong acid. Ethanoic acid is a **weak** acid.

The concentration of hydrogen ions is measured by pH.

pH numbers run from 0 to 14. Acids have pH from 0 to 6.
pH 7 is neutral. Alkali solutions have pH from 8 to 14.
Universal Indicator has a range of colours. They show pH.

> *When water and an acid mix*
> *The pH range is 0 to 6.*
> *If to alkali we change*
> *8 to 14 is the range.*

pH 0 1 2 3 4   5   6   7   8   9   10 11 12 13 14
    red   orange   yellow   green blue dark blue   violet
  → acid getting weaker → neutral → alkali getting stronger →

Experiment 12.4 To test the pH of solutions

a) Use 0.1 M solutions of hydrochloric acid and acetic
acid. Put 10 cm³ of the acid into a small beaker. Stir it
with a glass rod. Take a Universal Indicator paper.
Touch it with the wet rod. Record the pH.

Add two drops of M sodium hydroxide solution. Stir
with the rod. Touch the paper. Write down the pH.
Add two more drops. Stir. Find the pH of the liquid. Go
on doing this. Draw a graph of pH against drops.

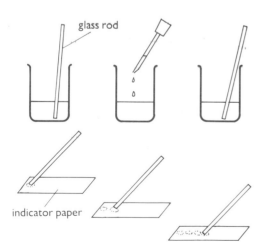

glass rod

indicator paper

b) Take solutions of as many substances as possible. Use
drops of Universal Indicator to test their pH. Salt, lime
water, aqueous ammonia, carbon dioxide, magnesium
oxide in water, 'bicarbonate of soda', copper(II)
sulphate, ammonium chloride, detergent can all be used.

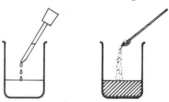

c) Put 10 cm³ of M hydrochloric acid into a small
beaker. Add two drops of Universal Indicator. Drop in a
small pinch of sodium hydrogencarbonate. Stir with a
glass rod. Add another pinch. Stir. Repeat this. After
each addition note the final pH.

> What does the graph show?
> Is each solution in b) acid, alkaline or neutral?
> Is it strongly acid or alkaline?
> What is 'bicarbonate of soda' used for?

The graphs show neutralization of acid by alkali. At
neutral there is a large change of pH.

Some salt solutions are not neutral.

Energy from chemical change

We have found out *how* elements combine. We also know the main reason *why* they do this. By combining they reach a lower energy level. The compound formed has less energy than the elements. They have fallen down the energy ladder.

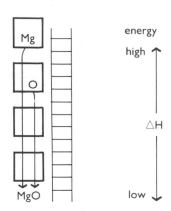

The energy they lose appears mainly as heat. Other reactions happen for the same reasons. The products have less energy than the reactants. This means that nearly all reactions give heat. If they do, they are **exothermic reactions.**

We use heat from exothermic reactions. It warms the rooms we live in. We use it to get hot food and hot water. In power stations it is turned into electrical energy. Measuring this heat is important.

How do we measure heat? We can only measure it by its effects. Heat makes things get hotter.

The unit of heat or energy is the joule, J.
Make 1 gram of water rise in temperature 1°C. This needs 4.2 joules. One joule is a very small amount. The kilojoule is a more useful unit.
   One kilojoule (kJ) = 1000 joules.

To measure heat: Measure out some water. Put the heat into it. Measure its rise in temperature.

We shall now try to measure the heat from a reaction.

Experiment 13.1 To measure the heat from burning

a) Fill a spirit lamp with ethanol. Cover the wick with a cap. Measure the mass of the lamp. With a measuring cylinder put 250 cm^3 of water into a metal can. 250 cm^3 of water has a mass of 250 grams.

spirit lamp

Stir the water. Take its temperature. Support the can in a stand. Put the lamp under it. Remove the cap. Light the wick. Let the flame heat the can.

Stir the water steadily. Let its temperature rise about 15°C. Remove the lamp. Quickly put the cap over the wick. Measure the new mass of the lamp. Go on stirring the water. Note its highest temperature. Work out how much heat has entered the water.

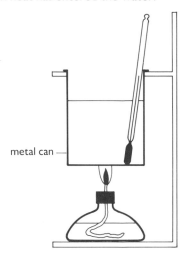
metal can

Ethanol burns in air. The reaction gives heat.
Does all this heat go into the water?
Where else could some of it finish up?
Does the result you worked out include this heat?
Is this result the exact heat given by burning?
How much ethanol did you burn?

In my result the temperature of the water rose 16°C. The mass of the lamp got less. It fell by 0.8 g.

To raise the temperature of 1 g of water 1°C needs 4.2 J. 250 g of water need 250 times as much heat as 1 g. The temperature of the water went up by 16°C. This would need 16 times as much heat as for 1°.

250×4.2 J

$250 \times 4.2 \times 16$ J $= 16800$ J $= 16.8$ kJ

This is the heat which went into the water. The metal can also became hotter. It took in heat. The air round the flame got hotter. This uses heat. The experiment takes time. Water and can are hot. Both lose heat to the air. None of this heat ends up in the water. So none of it is measured.

That means our result is only a rough one. We can make it more accurate. We can cut down heat losses. This means that more of the heat gets into the water. We can measure the heat used by the metal can.

We need to compare the heat given by different fuels. How much of each fuel shall we use? 1 g? 1 kg? 1 mole?

Burning is a reaction. It is the fuel molecules which react. We should use the same number of molecules of each fuel. One mole of each fuel contains the same number. This number is L. It is Avogadro's number.

$C_2 \quad H_6 \quad O$
$24 + 6 + 16 = 46$

$H = 1, C = 12, O = 16.$

The mass of the lamp gets less. The loss is the mass of ethanol which burnt. It was 0.8g. The formula of ethanol is C_2H_6O. One mole of it has a mass of 46 grams.

0.8 g of ethanol burning gives 16.8 kJ.

46 g of ethanol burning gives $\dfrac{16.8 \times 46}{0.8} = 966$ kJ

Burning is combustion. This heat quantity is called the **heat of combustion of ethanol**. We did not measure all the heat. An accurate value is much higher than this.

Ethanol belongs to a family of compounds. Measure the heat of combustion of some others. Methanol is CH_4O. Propanol is C_3H_8O. Butanol is $C_4H_{10}O$.

*b) Use the apparatus shown in the photograph. The vessel contains water. Measure its volume. The pump draws air through. The fuel burns in it.

The hot air goes through the spiral. It gives up its heat. This passes into the water. The water is stirred. Its temperature rise is noted.

The spirit lamp burns liquid fuel. Solid fuel burns in the cup. Both burn 'inside' the water. Loss of heat is small. Heat taken by the apparatus is measured.

This is the **heat of combustion** of one mole

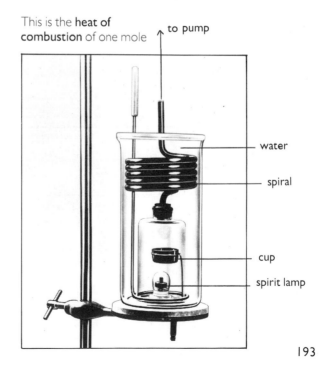

to pump

water

spiral

cup

spirit lamp

Energy from other reactions

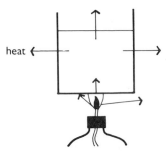

The burning fuel gives heat. Some of it goes into the water. Some is used in heating other things, such as the can. We measure only the heat which ends up in the water. Our result was not accurate. We must try to measure *all* the heat from a reaction.

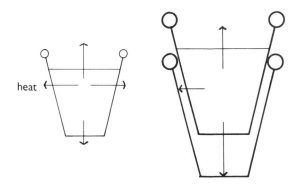

The can	We can find out how much heat it takes in. Or we can use a different vessel. A thin plastic cup uses almost no heat.
Thermometer	This gets hotter. The heat it uses is very small. It hardly affects the result.
The air	Some heat from the flame warms the air. The water in the can becomes warm. Warm water cools. It does so by losing heat to the air. We can reduce both losses. Do the reaction inside the cup. Stand the cup in another one.

Acid neutralizes alkali. This reaction gives heat. We shall try to measure this heat accurately. We shall use molar solutions of the alkali and acid.

Experiment 13.2 To measure the heat of neutralization

Measure acid and alkali with measuring cylinders. Put 50 cm^3 of M hydrochloric acid in a plastic cup. Stand this cup in an empty one. Put 50 cm^3 of M sodium hydroxide solution in another plastic cup.

Take the temperature of each solution. Use thermometers marked in tenths of a degree. Stir the acid. Pour in the alkali. Record the temperature of the mixture.

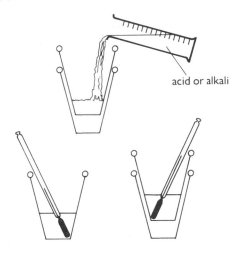

acid or alkali

Is the reaction exothermic?
What is the rise in temperature of the acid?
How much does the alkali temperature rise?
Work out how much energy the reaction gives.

Acid and alkali react as soon as they mix. The reaction is exothermic. The heat is produced *inside the liquid*. It reaches its highest temperature at once.

The plastic cup gets warmer too. It uses very little of the heat. There is an air space between the cups. Very little heat is lost to the outside air. Nearly all the heat stays in the water and is measured.

We can take 1 cm^3 of each liquid as one gram. One gram of each needs the same heat as 1 g of water.

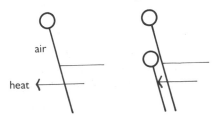

Working out the result

In one experiment acid and alkali were at 16°C. The mixture reached a temperature of 22.8°C. The rise in temperature of both liquids was 6.8°C. Altogether 100 g of water rose in temperature 6.8°C.

1 g of water rising 1°C in temperature needs 4.2 J.
100 g rising 1°C in temperature needs 100 × 4.2 J.   420 joules
100 g rising 6.8°C in temperature needs 420 × 6.8 J.   2860 joules
To avoid large numbers we write this as 2.86 kJ.   2.86 kilojoules

Neutralizing 50 cm³ of M acid gives 2.86 kJ of energy.
Neutralizing 1000 cm³ would give twenty times as much.   57.2 kJ
1000 cm³ of M solution contains one mole of acid.
Neutralizing one mole of acid gives 57.2 kJ of heat.

From this reaction *we gain* 57.2 kJ of heat energy.
The *substances which react give up*, or lose it.   −57.2 kJ
So heat from an exothermic reaction is given a minus sign.
When moles of a substance react, we use the symbol ΔH. ΔH is delta H

$$HCl + NaOH \rightarrow NaCl + H_2O \qquad \Delta H = -57.2 \text{ kJ}$$

This equation shows the result. It does not show how it happens. The acid is in solution as ions H^+ and Cl^-. Sodium hydroxide is a solid ionic compound. Its solution contains the ions Na^+ and OH^-.

H^+ and OH^- ions combine. They form molecules of water. This is because water is a covalent compound.

Na^+ and Cl^- ions remain in the solution. If the water evaporates, crystals of salt will form. Neutralization of acid by alkali is summed up by

$$H^+(aq) + OH^-(aq) \rightarrow H_2O(l) \quad \Delta H = -57.2 \text{ kJ}$$

$H^+ + Cl^-$ acid
| |
$OH^- + Na^+$ alkali
↓   ↓

H_2O
        $Na^+ + Cl^-$

salt solution
↓
$Na^+ Cl^- Na^+ Cl^-$
$Cl^- Na^+ Cl^- Na^+$
crystal

Experiment 13.3 Other acids and alkalis

Use other acids in place of hydrochloric acid. Put 50 cm³ of M acid solution into a plastic cup. Stand it in an empty one. Use ethanoic (acetic) acid, nitric acid and sulphuric acid.

Add to each 50 cm³ of M sodium hydroxide solution. Take the temperatures of acid, alkali and mixture. Work out ΔH for each mole of acid neutralized.

Are the results for all four acids the same?

Displacement reactions

Do you remember the Activity Series? It shows the metals in order of how active they are. The most active is at the top. The bottom metal is the least active. The Series also includes hydrogen.

Each metal is more active than those below it. It can push these metals out of their compounds.

In Book I we put zinc into copper(II) sulphate solution. Copper formed on the surface of the zinc. Zinc displaces copper. It is more active. It will appear above copper in the Activity Series.

We shall find out if this reaction gives heat. If so, we shall measure it. Will the result be accurate? Only if we avoid heat losses.

potassium
sodium
calcium
magnesium
aluminium
zinc
iron
lead
hydrogen
copper
silver
gold

Activity Series

Experiment 13.4 Heat from a displacement reaction

Measure 25 cm^3 of 0.5 M copper(II) sulphate solution. Put it into a plastic cup. Stand this in another cup. Measure out about 1.5 g of zinc powder.

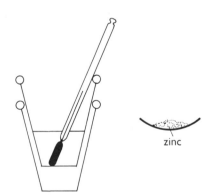
zinc

Work in pairs. Start a stop clock. Stir the liquid steadily. Take its temperature every half minute. When it is constant, record it.

At the next half minute exactly, pour in the zinc. Keep stirring. Take the temperature each half minute for four or five minutes.

> Record the results as a table.
> Use them to plot a graph.
> Is the reaction exothermic?

Reaction takes a little time to complete. It is exothermic. The graph shows how temperature changes.
AB shows the rapid rise caused by the reaction.
BC shows the hot solution cooling.
The dotted line CD shows how this would go on.
E shows what the temperature would have been
if reaction had taken place instantly
and no heat had been lost from the solution.

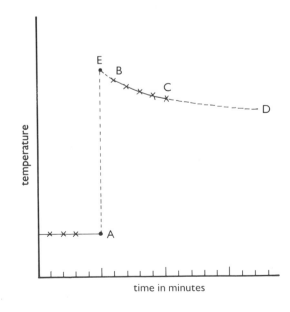

To raise the temperature of 1 g of water 1 °C needs 4.2 joules.
25 cm^3 of solution needs the same heat as 25 g of water.

The solution starts at temperature A, say 16°C.
It would have risen to temperature E, say 40°C.

25 g of water rising 24°C takes in $25 \times 24 \times 4.2$ joules.

2520 joules = 2.52 kJ

We need to know the heat given by one mole.
25 cm^3 of 0.5 M solution gives 2.52 kJ.
1000 cm^3 of solution would give 40 times as much.

$40 \times 2.52 = 100.8$ kJ

The solution is 0.5 M. 1000 cm^3 contains 0.5 mole.
One mole gives twice as much heat as 0.5 mole.

$2 \times 100.8 = 202$ kJ

This energy is given up by the reactants. They are losing energy. This energy is given a minus sign.

$\Delta H = -202$ kJ

The equation is
$$Zn(s) + Cu^{2+}(aq) + SO_4^{2-}(aq) \rightarrow Zn^{2+}(aq) + SO_4^{2-}(aq) + Cu(s)$$
zinc + copper sulphate solution $\rightarrow$ zinc sulphate solution + copper

The sulphate ions do not change. They are there all the time.
So the change is

$$Zn(s) + Cu^{2+}(aq) \rightarrow Zn^{2+}(aq) + Cu(s)$$

Zinc is better at turning into ions than copper is.
This is what being more active means.

Experiment 13.5 Other displacements

Repeat Experiment 13.4. Use aluminium, iron and lead powders with copper(II) sulphate solution.

Compare the heats of displacement. Can they be used to place the metals in the Activity Series?

Things to do

1. Measure heats of reaction for metals and dilute acid. Use 25 cm^3 of M hydrochloric acid. Drop in 1 g of metal. Place the metals in order.

2. Find out if there are energy changes when a substance dissolves. Put 25 cm^3 of water in a cup. Stir in 1 g of substance. Record any change in temperature. Work out the change for one mole of dissolved substance. Are any changes endothermic?

Summary

In chemical changes new substances form. They usually have less energy than the reactants. The energy difference is set free. It appears in most cases as heat. Light, sound and electrical energy can appear.

Most reactions give out heat. They are exothermic. A few take in heat. They are endothermic. The energy change is shown as ΔH. It is $-$ for exothermic and $+$ for endothermic reactions.

We use heat of combustion to compare the heating value of fuels. Heat of displacement can be used to place metals in the Activity Series.

Chemical energy → electrical energy

In Experiment 13.4 copper sulphate solution and zinc react. Zinc sulphate solution and copper are formed.

$$Zn(s) + Cu^{2+}(aq) + SO_4^{2-}(aq) \rightarrow Zn^{2+}(aq) + SO_4^{2-}(aq) + Cu(s)$$

The sulphate ions take no part. The real change is

$$Zn(s) + Cu^{2+}(aq) \rightarrow Zn^{2+}(aq) + Cu(s) \quad \Delta H = -202 \text{ kJ}$$

The reaction is exothermic. Chemical energy is being turned into heat. It can be turned into electrical energy instead. The device for doing this is a cell.

The Daniell cell makes use of this reaction. It is named after its inventor. A copper can holds copper(II) sulphate solution. A porous pot stands in it. It holds a zinc rod in zinc sulphate solution. We can make and test this cell.

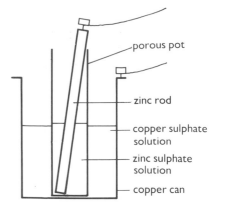

porous pot

zinc rod

copper sulphate solution

zinc sulphate solution

copper can

Experiment 13.6 Testing simple cells

Put two 100 cm^3 beakers side by side. Half fill one with copper (II) sulphate solution. Support in it a strip of copper foil. This makes a half cell.

Make a second half cell in the other beaker. Use zinc foil in zinc sulphate solution. Wet a strip of filter paper with dilute sulphuric acid. Put one end of it in each solution.

Connect each metal to one terminal of a voltmeter. Use wires joined to crocodile clips. If no reading shows, reverse the connections. Note which metal is joined to the + terminal. Record the highest voltage shown by the cell.

Set up other half cells. Hold the clean metal in a solution of its sulphate. Connect each in turn to a copper half cell. Use filter paper dipped in dilute acid. Record the voltage for each full cell. Use magnesium, iron and lead as suitable metals.

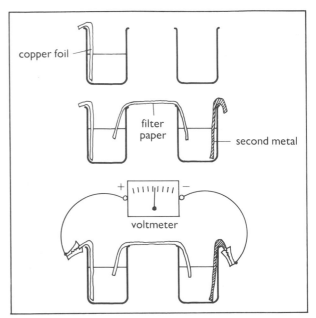

copper foil

filter paper

second metal

voltmeter

Two half cells joined in this way make a complete cell.
Which metal is the positive pole of each cell?
Which metal gives the highest voltage?
List the metals in order of voltage size.
Does the voltage shown by each cell change?

Cu^{2+} is pronounced C U two plus. The 2 explains why we call its compounds copper(II) compounds.

They put Daniel in the den of lions
The lions gave him L.
So Daniell used the den of ions
To make the first electric cell.

These cells provide electric current. How do they work? They are easy to set up. What disadvantages do they have?

Take the Daniell cell. Zinc atoms form ions, Zn^{2+}. Each atom loses two electrons. *These electrons remain on the zinc.* They make it the *negative pole* of the cell. The zinc ions pass into solution.

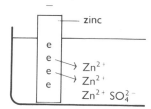

Copper atoms ionize less easily than zinc atoms. In fact, copper ions leave the solution. They join the copper plate. It becomes the positive pole.

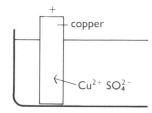

A voltmeter is a conductor. It connects copper and zinc. Electrons flow through it from zinc to copper. This flow of electrons is the current. More zinc ionizes to replace them. The zinc slowly dissolves.

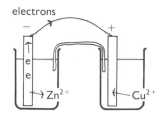

The reaction which provides energy is in two parts. One half cell gives electrons

$$Zn(s) \rightarrow Zn^{2+}(aq) + 2e$$

The other half cell uses them

$$Cu^{2+}(aq) + 2e \rightarrow Cu(s)$$

Copper is deposited on the copper plate.

Solutions in the two half cells must be connected. In the Daniell cell the pot is porous. In the experiments we used filter paper wet with acid.

Magnesium is more active than zinc. More of its atoms ionize. It will have a *bigger negative charge* than zinc. Its half cell gives a bigger voltage. Voltages fit in with Activity Series position.

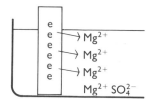

These cells have three disadvantages. The solutions in them make them awkward to use. They fail when a substance in them is used up. The voltage drops quickly because hydrogen may form.

The dry cell

This is what we call a 'battery'. It avoids some difficulties. Solutions are replaced by a wet paste. The battery's negative pole is a zinc pot. The zinc ionizes giving electrons. It is slowly used up. Zinc ions pass into the wet paste. The pot is slowly used up. It reacts even when the cell is not in use. Paste leaks out through the holes.

The positive pole is the carbon rod. Electrons flow to it. Manganese(IV) oxide oxidizes any hydrogen to water. The cell fails when it is completely used up.

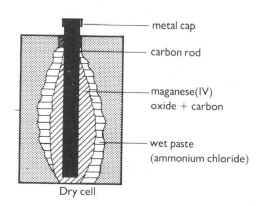

Dry cell

Cells – primary and secondary

A cell makes ions and electrons move. The force which moves them is called the e.m.f. of the cell. E.m.f. stands for electromotive force. It is measured in volts, V. In a solid, current is a flow of electrons. In a solution it is a flow of ions.

A battery is a set of cells joined in series. This means the negative pole of one cell is joined to the positive pole of the next. The e.m.f. of a dry cell is 1.5 volts. A battery of four cells has an e.m.f. of 6 V.

In the cell substances react. When they are used the cell is dead. We cannot reverse these chemical changes. This kind of cell is a primary cell. In some cells we can reverse the changes. These are called secondary cells. We shall make and test one.

Repeat the electrolysis of dilute sulphuric acid, Experiment 4.1. What gases are formed? Do they react with platinum electrodes? Will they react with lead?

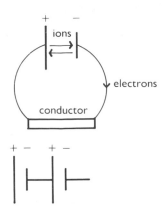

> Judge: What are the prisoners' names?
> Constable: Anne Ode and Cath Ode, your Honour.
> Judge: Have they been charged?
> Constable: Yes, your Honour, with assault and battery.
> Judge: Put them in a dry cell.

Experiment 13.7 To make a lead cell

Clean two strips of lead foil. Support them in 4 M sulphuric acid. Connect them to a six-volt power pack. Note which strip is the positive anode.

Switch on. Let the current run for five minutes. Disconnect the pack. Connect a bulb across the electrodes. Use the power pack again. Run the current for ten minutes. Test with the bulb again.

> Which lead electrode changes and in what way?
> The bulb lights up then dims. What does this show?
> Does the bulb light up for longer the second time?

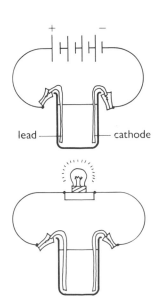

At the anode oxygen is set free. It reacts with the lead. Brown lead(IV) oxide forms.

$$Pb(s) + O_2(g) \rightarrow PbO_2(s)$$

At the cathode hydrogen forms. It does not react with lead.

The bulb shows that current flows. We have made a cell. It is the lead-acid accumulator. Its e.m.f. is 2.05 V. The bulb dims. The cell fails as the brown oxide is used up. Passing a current reverses this. The cell is charged again.

The lead–acid accumulator

Its negative pole, or electrode, is spongy lead. It provides electrons: $Pb(s) \rightarrow Pb^{2+}(aq) + 2e$
Its positive pole is brown lead(IV) oxide (lead 4). It uses electrons: $Pb^{4+}(s) + 2e \rightarrow Pb^{2+}(s)$

As current flows these changes continue. Pb^{2+} ions at both poles form white lead(II) sulphate. This uses sulphuric acid. The acid in the cell becomes weaker. The cell will give no current. It is 'flat'.

The changes can be reversed. Current is passed through in the reverse direction. The dilute sulphuric acid is electrolyzed. The results are:

At the cathode, hydrogen forms. It reduces lead sulphate. The cathode becomes lead metal again.

At the anode oxygen forms. It oxidizes the lead sulphate. The anode becomes lead(IV) oxide again.

Both changes produce sulphuric acid. The acid concentration rises. The electrodes are renewed. The e.m.f. is 2.05 V again. The gases no longer react. They appear as bubbles. The cell 'gasses'. This shows that the cell is 'charged'.

Summary

Cells provide electrical energy, or current. It comes from chemical change. A cell has two poles or electrodes. They stand in solutions of ions.

The negative pole gives electrons. They come from a metal turning into ions. $Zn(s) \rightarrow Zn^{2+}(aq) + 2e$

They flow through any conductor connecting the poles. This flow of electrons is the current. The positive pole uses the electrons for a chemical change. Current in the solution is a flow of ions.

Electrical pressure in the cell causes these flows. It is the e.m.f. of the cell, measured in volts. Chemical changes in some cells cannot be reversed. The cell is dead when the substances are used up. This kind of cell is a primary cell.

In secondary cells the changes can be reversed. The cells can be 'charged'. The lead accumulator has a negative pole made of lead. Its positive pole is lead(IV) oxide. The poles stand in sulphuric acid.

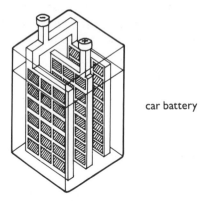

car battery

six cells in series = 12 volt battery

$$PbSO_4 + H_2 \rightarrow Pb + H_2SO_4$$

$$2PbSO_4 + O_2 + 2H_2O \rightarrow 2PbO_2 + 2H_2SO_4$$

Questions

1. Glucose is a sugar, $C_6H_{12}O_6$. It burns in oxygen. Relative atomic masses: $C = 12, H = 1, O = 16$.
   $C_6H_{12}O_6(s) + 6O_2(g) \rightarrow 6CO_2(g) + 6H_2O(g)$
   $\Delta H = -2800$ kJ.
   What does ΔH stand for?
   What does kJ mean?
   What does the minus sign in -2800 kJ mean?
   How many grams of glucose make one mole?
   How much heat is given by burning one mole of it?
   Glucose is a food. How does it give us energy?
   Plants make glucose from carbon dioxide and water. Is this reaction exothermic or endothermic?
   Where does the energy stored in glucose come from?
   In sunshine, pond weed gives bubbles of gas. Which gas is it? Write an equation for the reaction.

2. A metal can displace other metals from solutions of their salts. Describe an experiment to determine the heat of reaction. The results can be used to place the metals in the Activity Series. How?

3. You are given zinc and copper metals and solutions of their sulphates. Say how you would use them to make a cell.
   How would you measure its e.m.f.?
   What is a primary cell?
   How is it different from a secondary cell?
   What materials make up a lead–acid cell?
   How many make a 12 volt car battery?

Carbon compounds

Carbon is an element. We studied it in Section 8. Two forms of carbon occur in nature. They are diamond and graphite. Both are also manufactured.

Diamond is very hard. It is transparent, like glass. Graphite is soft and grey. If they are pure, both contain only carbon atoms. Both are crystalline solids. But they have different crystal shapes.

Then why are they so different? Because their carbon atoms are packed differently. Different forms of the same element are allotropes. They make up the **allotropy** of carbon.

We have come across many carbon compounds. Foods contain them. So do many minerals. Large numbers are present in crude oil. Air contains carbon dioxide. Fuels are carbon compounds.

A carbon atom can link up with atoms of other elements. It can also link to other carbon atoms.

The atomic number of carbon is 6. Its atom will contain six electrons. Two fill the first shell. The other four must be in the second shell. The electron structure of a carbon atom is 2.4.

The nearest noble gases are helium and neon. Their structures are 2 and 2.8. To be like helium the carbon atom must lose four electrons. This is impossible. Can you see why?
To be like neon it must gain four. It can do this only by sharing. Most carbon compounds are covalent.

A hydrogen atom has one electron. It can be used in one shared pair. A carbon atom needs four. It forms shared pairs with four hydrogen atoms. The compound is methane. Its formula is CH_4.

The electron structure of oxygen is 2.6. Its atom can form two shared pairs. A carbon atom needs to share four pairs. It joins two oxygen atoms. The formula of carbon dioxide is CO_2. It is covalent.

Carbon dioxide molecules are small. They are covalent. This explains why it is a gas. Methane has small covalent molecules. It is also a gas. Natural gas is chiefly methane. We also get it from crude oil. What are its properties?

These diamonds are one allotrope of carbon . . .

. . . this graphite is another.

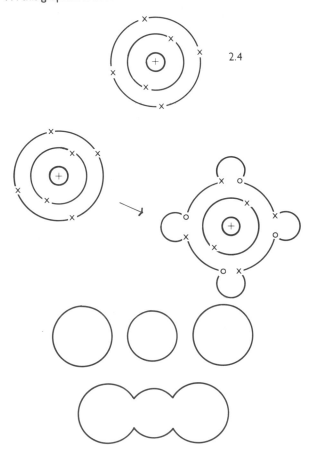

2.4

Experiment 14.1 The properties of methane

Grind soda lime and anhydrous sodium ethanoate (acetate) to powder. Heat them separately in an oven at 200°C. Mix well equal parts of each.

Heat this mixture. Collect the gas in large test tubes. Stopper them. Use one for each test.

Test methane with a lighted splint. Hold a large test tube over the flame. Add lime water to this tube. Stopper it and shake the solution.

Shake methane with water. Add a litmus paper. Put in drops of Universal Indicator. Shake methane with a) acidified manganate(VII) solution and b) a solution of bromine in water.

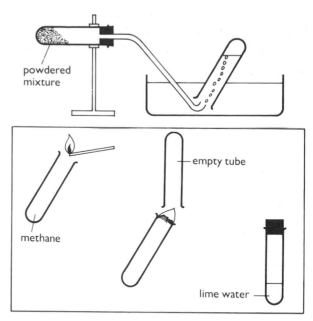

powdered mixture

empty tube

methane

lime water

Apart from burning, does methane react?

Methane burns at the mouth of the tube. The flame is like a bunsen flame with its air hole closed. Carbon dioxide and water form in the tube over the flame.

Methane does not react with water, acids or alkalis. It is not oxidized or reduced. The brown bromine colour fades. Methane reacts with bromine slowly.

A carbon atom has four covalent bonds. In methane they are filled by hydrogen atoms. The molecule can change in only one way. A hydrogen atom can be taken away. Another atom can go in its place.

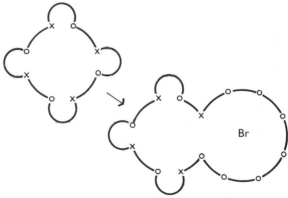

Br

This kind of change is called **substitution**. It can happen to all four hydrogen atoms. A compound which can react only by substitution is **saturated**.

$$CH_4 + Br_2 \rightarrow CH_3Br + HBr$$

CH_3Br is bromomethane.

Methane contains only carbon and hydrogen. It is a **hydrocarbon.** Hydrocarbons burn in air. The usual products are carbon dioxide and water.

Two carbon atoms can join by a covalent bond. Each provides one electron of the shared pair. Each now has five electrons in its outside shell. Both need three more. Each can link to three hydrogen atoms. This substance is a hydrocarbon too.

Like methane it is a gas. It has no smell or colour. It reacts only by substitution and by burning. It is saturated. It is called ethane, formula C_2H_6.

Methane and ethane are very much alike.

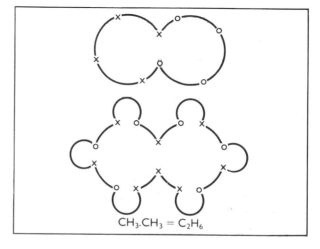

$CH_3.CH_3 = C_2H_6$

Formula and family

Plants and animals are living 'organisms'. Many carbon compounds come from them. People once thought that they all did. They called the study of carbon compounds 'Organic Chemistry'. Over two million of them are known. Most of them are manufactured. This makes 'Carbon Chemistry' a better name.

Each compound has a formula. There are four ways of writing it. We can show how using methane and ethane.

Methane		**Ethane**
CH_4	This is the molecular formula. Use it in equations.	C_2H_6

These drawings show the shared pairs of electrons.

H | H-C-H | H	A shared pair is a covalent bond. Each bond is shown as a short line. The result is a structural formula. It shows which atoms are bonded together.	H H | | H-C-C-H | | H H
CH_4	A short form of this is often used. It shows which atoms are joined to each carbon atom.	$CH_3.CH_3$

We can think of each atom as a sphere. Hydrogen atoms are smaller than carbon atoms. They are roughly table tennis ball to tennis ball in size.

In methane four H atoms are evenly spaced round one C atom. Imagine four table tennis balls fixed to one tennis ball. Each table tennis ball is the same distance from the other three. Atomic models show this well. Remember that electron shells overlap.

In the drawings carbon atoms are black. Hydrogen atoms are white.

Three carbon atoms can join by covalent bonds. The middle one then has six electrons. It needs eight. It forms bonds with two hydrogen atoms.

Each end carbon atom now has five electrons. Each carbon atom can form bonds with three hydrogen atoms. The molecular formula is C_3H_8. The structural formula is shown in both ways.

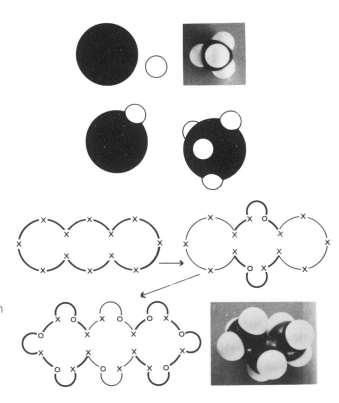

204

C_3H_8 is a hydrocarbon and a gas. It burns forming carbon dioxide and water. Apart from this it reacts by substitution only. It is very much like ethane and methane. It is called propane (pro-pane).

All three belong to the same family. This kind of family is called an 'homologous series'. This one is the alkanes. Their names are in two parts. The first part says how many carbon atoms. The second part is the family name. -ane stands for alkane.

Look at their formulae. The next one has four carbon atoms. How many H atoms in the molecule?

This is butane (bewt-ane). It has a chain of four carbon atoms. They are bonded to ten hydrogen atoms. Butane is a gas which burns and reacts by substitution. Its formula is C_4H_{10}. The next alkane is C_5H_{12}.

Four carbon atoms can be linked in a different way. We can put three in a chain. We can join the fourth to the middle one. Ten hydrogen atoms are needed. This means there are two compounds with molecular formula C_4H_{10}. Such different compounds are called isomers.

meth- means one carbon atom
eth- means two
prop- means three
but- means four
pent- means five

$$CH_4 \quad C_2H_6 \quad C_3H_8 \quad C_4H_?$$

```
-C-C-C-C-
 H H H H
 | | | |
H-C-C-C-C-H
 | | | |
 H H H H
```

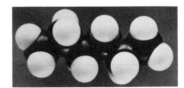

Which one is this?

```
      |
     -C-
      |
-C-C-C-
      |
```

Summary

A carbon atom can form four covalent bonds. They can be filled by four hydrogen atoms. This compound is called methane. Its formula is CH_4.

Methane is a gas hydrocarbon. It burns. It can react in only one other way. Its hydrogen atoms can be replaced by other atoms. This is substitution. Substances which react in this way are saturated.

A covalent bond can link two carbon atoms. Each can then link to three H atoms. The compound is ethane. Its formula is C_2H_6 or $CH_3.CH_3$. Methane and ethane belong to a family or homologous series. They are alkanes. The first part of the name shows how many carbon atoms. The ending is the family name, -ane.

A structural formula shows which atoms are joined. A short line stands for a covalent bond.

Compounds can have the same molecular formula but different structural formulae. Such compounds are called isomers.

Questions

1. Write down the formulae and names of the first four alkanes. Give the name and formula of the fifth.

2. What is meant by
   a) an homologous series,
   b) substitution reaction,
   c) hydrocarbon,
   d) isomers,
   e) a saturated compound,
   f) a covalent bond,
   g) a structural formula.

Alcohols

For thousands of years people have made wine from grapes. Crushed grapes give a sweet juice. The skins carry a fungus like yeast. It helps to break down the sugar in the juice. A gas is formed. The mixture froths. The reaction is called fermentation.

Yeast will ferment sugar in the same way.

Picking grapes to make wine.

*Experiment 14.2 New substances from fermentation

Dissolve 20 g of sugar in 80 cm³ of water. Gently grind 10 g of yeast in 20 cm³ of water. Pour both into a flask. Add crystals of potassium nitrate, ammonium phosphate(V) and magnesium sulphate.

Warm the liquid to about 25 °C. Plug the flask loosely with cotton wool. Put it in a warm place. When it froths, bubble the gas through lime water. When frothing stops, carefully pour off the liquid.

Distil the liquid as shown. Collect the distillate. Stop when the temperature rises above 80 °C. Smell the liquid. Pour a little of it on a watch glass. Touch it with the flame of a lighted splint.

Put ½ cm depth of it into each of four tubes.
*To 1 add a scrap of sodium. Test for hydrogen.
*To 2 add a pinch of phosphorus(V) chloride. Hold a damp litmus paper in the mouth of the tube.
Take potassium manganate(VII) solution. Add dilute sulphuric acid. Add drops of this solution to tube 3.
Acidify potassium dichromate(VI) solution. Add drops of it to test tube 4. Warm both test tubes.

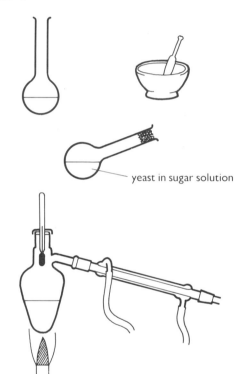
yeast in sugar solution

> Fermentation gives carbon dioxide. A colourless liquid distils over.
> What is its boiling point?
> Did you recognize its smell?

The colourless liquid smells like 'meths'. It boils at 78 °C. It has always been called 'alcohol'. Experiments prove that its formula is C_2H_6O.

Wine, beer and other drinks are made by fermentation. They all contain this substance. This is why they are called alcoholic drinks.
Yeast consists of living cells. They contain substances called enzymes. They are catalysts for the breakdown of sugar molecules.

What gases are formed in tests 1) and 2)?

$$C_{12}H_{22}O_{11} + H_2O \xrightarrow{\text{invertase}} C_6H_{12}O_6 + C_6H_{12}O_6$$

sucrose   glucose   fructose

$$C_6H_{12}O_6 \xrightarrow{\text{zymase}} 2C_2H_6O + 2CO_2$$

Sucrose is ordinary sugar. Glucose and fructose are sugars too.
Invertase and zymase are two of the enzymes in yeast.

C_2H_6O is a liquid. It has a low boiling point. It is the liquid in 'meths'.
We know that meths does not conduct a current. These are covalent
properties. The atoms must be joined by covalent bonds. Each bond is
shown as a line.
The structural formula of C_2H_6O is ———

The −O−H part is called a hydroxyl group. Apart from this the
molecule is like ethane. Its name begins with ethan-. The ending-ol
shows an −O−H. The correct name for 'alcohol' is ethanol.

$CH_3.CH_2OH$

this is ethanol

Ethane is not reactive. Ethanol is. The −OH group is the only
difference between them. It must be the active part of ethanol.
It reacts with the metal sodium.
Metals displace hydrogen from the −OH group.

Ethanol reacts with phosphorus chlorides. A misty gas turns litmus red.
It must be hydrogen chloride.
The hydroxyl group is replaced by a chlorine atom.

chloroethane

Ethanol reacts with purple potassium manganate(VII) solution. The
products are colourless. Orange dichromate(VI) solution changes to
green. What do these colour changes tell us? What has happened to
the ethanol?

The colour changes show that oxygen is being given up. Ethanol is
oxidized in two stages.
The first stage forms ethanal. It is an aldehyde.
Stage 2 forms an acid. It is ethanoic (acetic) acid.

The aldehyde group is $-C\overset{\displaystyle H}{\underset{\displaystyle O}{\big<}}$. The acid group is $-C\overset{\displaystyle O-H}{\underset{\displaystyle O}{\big<}}$

Both contain a carbon atom linked to an oxygen atom. This needs two
covalent bonds. Two bonds linking two atoms are called a **double
bond.**

Like ethane, ethanol burns. The products are carbon dioxide and
water. As meths, ethanol is a common fuel.

C = O

Fermentation, food and drink

Ethanol in everyday life

Wine comes mainly from grapes. At one time they were trampled on to get the juice. Now modern presses are used. The juice ferments for about a week. The wine is then run off into vats.

Here it becomes clear and matures. Then it is bottled. Dark-skinned grapes make the wine red in colour. White or yellow grapes give a white wine.

Sugar is added to some wines before bottling. This will ferment inside the bottle. Carbon dioxide builds up inside. The bottle will open with a 'pop'. The wine is 'bubbly'. Winning motor racing drivers spray it about quite a lot!

Beer is made from barley. The grain is made wet and warm. It sprouts. Some of its starch is turned into sugar. Heating it in a kiln kills the little plants. The result is malted barley.

This malt is crushed in hot water and filtered. The liquid is called wort. It is boiled in copper pots with sugar and roasted hops. After cooling yeast is added. It ferments for a few days.

The yeast grows and multiplies. Some of it is removed. It is sold as 'brewers' yeast. Substances are added to clear the beer. Beer is bottled or put into casks. It contains carbon dioxide. This is what makes the froth or 'head' when it is poured.

Whisky is made from malt too. The kiln is heated by burning peat. The smoke from it gives flavour to the fermented liquid. The fermented liquid has about 14% ethanol.

This liquid is distilled in copper pots. The distillate has a higher percentage of ethanol in it. If this is distilled the percentage becomes bigger still. The result is whisky. It may stand for years before it is bottled. This process is called maturing.

Liquids made by distillation are called spirits. Brandy is made by distilling wine. Molasses can be fermented. Distilling the liquid gives rum. All spirits have a high percentage of ethanol.

Pressing grapes to extract the juice.

Whisky is distilled in large copper pots.

Grace's favourite boy friend Mike
Drank alcohol and crashed his bike.
He's lucky to be still alive;
If you drink you shouldn't drive

Alcoholic drinks contain ethanol. What does drinking it do to us?

Small amounts of alcohol make us more lively than usual. Our hearts beat faster. Alcohol sharpens our appetite for food. We worry less about what people think of us.

Larger amounts of alcohol slow us down. We react more slowly. Drivers are more likely to have accidents. They *think* they are still driving well. They risk other peoples' lives as well as their own.

Find out about breathalysers. The older model had crystals. Alcohol changed them from orange to green. Where have we met this change before? The new model shown has a meter. This shows the amount of alcohol breathed into it.

Constant drinking damages liver and kidneys. The habit is hard to break. Some people become alcoholics.

Mixing dried yeast for breadmaking.

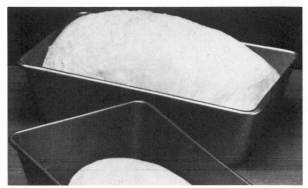
Loaves are left in a warm place to let the dough rise.

The new breathalyser test.

Bread is made using yeast. Flour and water are kneaded into dough. Yeast and sugar are mixed in. They ferment giving bubbles of carbon dioxide. This gas puffs up the dough, or 'raises' it.

The dough is shaped into loaves. They are baked in an oven. The carbon dioxide and ethanol escape. The high temperature kills the yeast. The bread is left full of holes. It is light and spongy in texture.

Vinegar is made from wine or beer. If wine or beer stand in air they become sour or sharp.
Ethanol reacts with oxygen. Ethanoic acid (acetic acid) is formed. Some bacteria speed up the change. Vinegar is a weak solution of the acid, CH_3COOH, in water.

Ethanol in industry

Experiment 9.2 gave a liquid. At most the liquid contains about 96% alcohol. It is called rectified spirit. Removing water from it gives absolute alcohol. Both are used as solvents. They dissolve many substances from paint to perfume.

Methylated spirit is rectified spirit with methanol. A violet dye may be added to it. It is used as a fuel. Methylated spirit is cheaper because less tax is paid on it. The methanol makes the alcohol poisonous to drink.

Alcohols and esters

There are many substances like ethanol. They belong to a homologous series. This series, or family, is the alcohols. Each has a hydroxyl group, $-OH$, in its molecule. Primary alcohols have a $-CH_2OH$ group.

All primary alcohols contain these two groups. This is why they are alike. They all have the properties of the two groups. The $-OH$ group reacts with sodium. All alcohols react with it giving off hydrogen.

Each $-OH$ group is bonded to a group of atoms. These groups are called radicals. They are methyl CH_3-, ethyl C_2H_5-, propyl C_3H_7-, butyl C_4H_9- and so on.

All alcohols react with acids.

H H
│ │
H-C-C-O-H   [radical]-OH
│ │
H H

[radical]-OH
    ↘
        [radical]-ONa + H

H H
│ │
H-C-C-O-H   is C_2H_5OH
│ │
H H

ethanol is ethyl alcohol

Experiment 14.3 Alcohol and acid

Put I cm depth of ethanol in a test tube. Add about ten drops of pure ethanoic acid. Put in ten drops of concentrated sulphuric acid. TAKE CARE! Warm the mixture gently. Smell it with great care.

Let it cool. Pour it into sodium carbonate solution in a small beaker. Smell it again with care.

> Is the smell pleasant?
> Do you recognize it?

The $-OH$ groups of acid and alcohol react. A molecule of water is formed. The sulphuric acid takes up this water. It also acts as a catalyst. The rest of the acid and alcohol molecules join to form an ester.

> acid + alcohol → ester + water

Flowers and fruit produce esters. When we smell them we breathe in ester vapour. So esters have a 'fruity smell'. Most natural fats and oils are esters.

$$CH_3C \overset{\displaystyle =O}{\underset{\displaystyle \diagdown O\text{-}H\, H}{}} \; \text{-}O\text{-}C_2H_5$$

acid   alcohol

↓

$CH_3COOC_2H_5 + H_2O$
ester   water

The names of isomers

We can use radical names to name isomers. Take C_4H_{10} again. Its atoms can be bonded in two ways. One has a chain of four carbon atoms. It is butane. Look at the formula. It shows the carbon atoms in a straight line. In the molecule they form a zigzag.

Take a chain of three carbon atoms. Put in the hydrogen atoms. The formula is C_3H_8. It is propane.

Take an H atom from the middle carbon. Put in a fourth carbon. This bonds to three H atoms. The formula is C_4H_{10}. It is an isomer of butane. It is a methyl radical in propane. It is called methylpropane.

H H H H
│ │ │ │
H-C-C-C-C-H
│ │ │ │
H-H-H-H

straight line

zigzag

H H H
│ │ │
H-C-C-C-H
│ │ │
H H H
propane

          H
          │
        H-C-H
H │ H
H-C-C-C-H
│ │ │
H H H
methylpropane
$CH_3.CH.(CH_3).CH_3$

There are two alcohols with formula C_3H_7OH.
Their structural formulae are shown.
Both have three carbon atoms.
Both are like propane. Both names begin with propan-.
Each has an $-OH$ group. Each name must end in -ol.

-C-C-C-

3 2 1

Number the carbon atoms from right to left.
A has its $-OH$ group on carbon 1. It is propan-1-ol.
B has its $-OH$ group on carbon 2. It is propan-2-ol.
Propan-3-ol is the same formula as propan-1-ol.

```
A      H H H                    B        H
       | | |                             |
    H-C-C-C-O-H                        H O H
       | | |                             | | |
       H H H                          H-C C-C-H
                                         | | |
                                         H H H
```

Summary

Alcohol usually means ethanol, C_2H_5OH. It is formed when yeast ferments sugar. The enzymes in yeast act as catalysts. They cause the breakdown of sugar. The frothing during reaction is carbon dioxide bubbles.

This is how alcoholic drinks are made. Distilling these gives spirits. They have a high ethanol content

Ethanol burns in air. It is a useful fuel. It is used in industry as a solvent. Ethanol is also used to make other important substances.

Rectified spirit is 95% ethanol and 5% water. Adding methanol, CH_3OH, makes it poisonous to drink. The mixture is called methylated spirit, or meths. A violet dye may be added as well. Pure ethanol is called absolute alcohol. It belongs to an homologous series called alcohols. They have hydroxyl, $-OH$ groups.

Sodium reacts with alcohols. It replaces the H atom of the $-OH$ group. Phosphorous chlorides replace the hydroxyl, $-OH$, group by a chlorine atom.
   alcohol + acid → ester + water.
Concentrated sulphuric acid acts as a catalyst and absorbs water.

Alcohols with a $-CH_2OH$ group are primary alcohols. They can be oxidized.
   $-CH_2OH + O \rightarrow -CHO + H_2O$
Substances with a $-CHO$ group are aldehydes.

Further oxidation gives acids. These acids all contain the group $-COOH$. CH_3COOH is ethanoic acid or acetic acid. Vinegar is a weak solution of it.

Questions

1. Fill in the blank spaces:

   Sugar is fermented by using _____. The frothing is caused by a gas. It is _____ _____. The other substance formed is 'alcohol'. Its formula is C_____. Its chemical name is _____. It belongs to a family or homologous _____.

   Yeast contains substances called _____. They break down sugar by acting as _____. The active part of ethanol is the _____ group, $-OH$. Metals such as _____ react with it. They replace the _____ of the $-OH$ group. The gas formed is _____. Phosphorus chlorides react with alcohols. The $-OH$ group is replaced by a _____ atom, $-$_____. Alcohol and acid react. They form an _____.

   Ethanol can be oxidized. The first product is an _____. Further oxidation gives an _____.

2. Write one sentence for each word to explain what it means:
   alkane, structural formula, primary alcohol, ester, substitution reaction.

Grace's science teacher taught her
Acid + base gives salt + water.
And another rhyme to test her
Acid + alcohol gives ester + water.

Breaking down other molecules

Sugar molecules are large. Fermentation breaks them down into smaller ones. One of the products is ethanol. We shall try to decompose ethanol.

sugar $\rightarrow$ ethanol $\rightarrow$?
$C_{12}H_{22}O_{11} \rightarrow C_2H_5OH \rightarrow$?

Experiment 14.4 To decompose ethanol

Put a 1 cm depth of ethanol into a test tube. Push rocksill in to soak up the liquid. Support the tube horizontally. Put a pile of porcelain bits in the middle of it. Broken evaporating dish will do.

Heat the porcelain fairly strongly. Collect the gas in large test tubes. Stopper them. Do the same tests on the gas as on methane, Experiment 14.1.

> What happens to a lighted splint in the gas?
> Does the gas burn forming carbon dioxide?
> Does it affect litmus or Universal Indicator?
> Does the gas react with bromine?
> Can potassium manganate(VII) solution oxidize the gas?

The broken porcelain acts as a catalyst. Heating it also heats the ethanol. So ethanol vapour passes over the catalyst. A colourless gas is collected. It burns with a yellow flame. The products are carbon dioxide and water. The gas is a hydrocarbon.

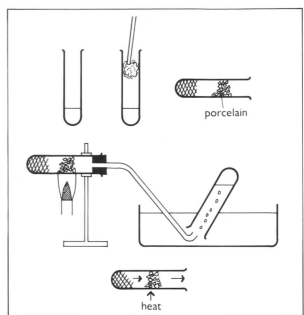

porcelain

heat

Its formula is C_2H_4. How are the atoms joined? Join the carbon atoms. Each forms 3 more bonds. C_2H_4 has four hydrogen atoms. Each carbon atom takes two. Each still has a spare electron. They form a shared pair. The carbon atoms are joined by two bonds. This linkage is called a **double bond**.

C_2H_4 has 2 carbon atoms. Its name will begin eth-. It is ethene. The ending -ene means a double bond. A double bond makes a molecule very reactive.

Ethene reacts with bromine. We know this because the brown colour of bromine disappears. A bromine atom adds to each carbon atom. The double bond changes to a single one. Two substances combine to form one. This is an addition reaction:

$$C_2H_4 + Br_2 \rightarrow C_2H_4Br_2$$

In an addition reaction, two or more molecules react to form only one product.

$C_2H_4Br_2$ is called dibromoethane. It could be made from ethane. Two H atoms of ethane are replaced by Br atoms. Two different compounds have this formula.

di bromo ethane

The one made from ethene has formula **A**. The other has both bromine atoms bonded to the same carbon. This is formula **B**. The two compounds are isomers. Their names must tell us which is which.

Number the two carbon atoms from the right. Use them to say where the bromine atoms are.

A
$$H-\overset{\overset{\displaystyle H}{|}}{C}-\overset{\overset{\displaystyle H}{|}}{\underset{\underset{\displaystyle Br}{|}}{C}}-H$$
$$\underset{Br}{}$$

B
$$H-\overset{\overset{\displaystyle H}{|}}{\underset{\underset{\displaystyle H}{|}}{C}}-\overset{\overset{\displaystyle H}{|}}{\underset{\underset{\displaystyle Br}{|}}{C}}-Br$$

$$-C-C-$$
$$2\ \ 1$$

A is 1,2-dibromoethane

B is 1,1-dibromoethane

In ethene, potassium manganate(VII) solution loses its purple colour. This shows it is losing oxygen. It is oxidizing ethene. The equation is shown.

$$\overset{H}{\underset{H}{>}}C=C\overset{H}{\underset{H}{<}} \quad + H_2O + O \quad \rightarrow \quad H-\overset{\overset{\displaystyle H}{|}}{\underset{\underset{\displaystyle OH}{|}}{C}}-\overset{\overset{\displaystyle H}{|}}{\underset{\underset{\displaystyle OH}{|}}{C}}-H$$

potassium manganate (VII) solution

This product has two —OH groups. It is an _____. It has two carbon atoms. Its name begins with _____. Each carbon atom carries an —OH group. Can you name the substance? (It is difficult!)

$$H-\overset{\overset{\displaystyle H}{|}}{\underset{\underset{\displaystyle OH}{|}}{C}}-\overset{\overset{\displaystyle H}{|}}{\underset{\underset{\displaystyle OH}{|}}{C}}-H \text{ is ethane-1,2-diol}$$

Alkenes are very reactive. They react by addition. Work out the reactions of ethene with hydrogen, hydrogen chloride HCl, hydrogen bromine HBr, chlorine Cl$_2$. Name the product formed in each case.

Compare ethane, C$_2$H$_6$ with ethene, C$_2$H$_4$.
Ethane is an alkane. It is saturated. It reacts by substitution.
Ethene is an alkene. It is unsaturated. It reacts by addition.

Ethanol molecules can be broken down. One product is ethene. It belongs to an homologous series called the alkenes.

$$C_2H_5OH \rightarrow C_2H_4 + H_2O$$

We have decomposed sugar. Sugar comes from plants. We shall now try a mineral substance. It is medicinal paraffin. It comes from crude oil or petroleum.

Experiment 14.5 Breaking down medicinal paraffin

Use the apparatus of the last experiment. Use the paraffin in place of ethanol. Use pumice in place of porcelain. Heat it strongly. Collect the gas.

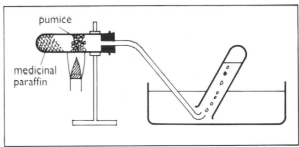

Is the gas likely to be a hydrocarbon?
What test will you use to find out?
Is the gas an alkane or an alkene?
What test will you use to find out?
What second test can be used to decide?
What suggests that the product has smaller molecules than the paraffin?

It burns giving carbon dioxide and water.
Add bromine solution. Alkenes react with it.
Use potassium manganate(VII) solution.
The product is a gas. The paraffin is liquid.

It could be a hydrocarbon.
Its brown colour goes.
It loses its purple colour.
Gases have small molecules.

213

Larger molecules to smaller molecules

We can break down large molecules by other methods.

On page 210 we made an ester. We used the reaction

acid + alcohol → ester + water

The ester has a bigger molecule than the acid or the alcohol. This is a condensation reaction.

In a condensation reaction two or more molecules react. One product is a molecule larger than either. The other is a simple molecule such as H_2O.

This reaction can be reversed:

ester + water → acid + alcohol

The reaction is very slow. It can be made faster by using a catalyst. Acids and alkalis act as catalysts. The large ester molecule breaks down to smaller ones.

Most natural oils and fats are esters. Castor oil is.

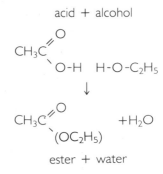

acid + alcohol

ester + water

Experiment 14.6 The breakdown of esters

a) Wear eyeshields. Put 2 cm³ of castor oil in a 50 cm³ beaker. Add with care 10 cm³ of 5 M sodium hydroxide solution. Keep it away from clothes or skin. It is very caustic.

Stir the mixture and gently heat it. Let it boil very gently for about five minutes. Add an equal volume of distilled water. Stir in half a spoonful of salt. Filter. Wash the solid with distilled water.

Half fill one test tube with distilled water. Half fill another with hard tap water. Add a lump of the solid to each. Stopper and shake the two tubes.

What happens to the castor oil?
How does the solid affect the two samples of water?
Do these results tell you what the solid is?

* b) Put a 1 cm depth of castor oil into a test tube. Add twice the volume of concentrated sulphuric acid. TAKE CARE! Stir the mixture with a glass rod. Pour it into distilled water in a beaker. Stir it. Carefully pour the water away. Oil is left in the beaker. Wash it again with distilled water.

Take test tubes of hard water and distilled water. Add a little oil to each. Shake the tubes.

How does the oil affect the two samples of water?
Are results for hard and soft water different?
Does this tell you what the oil is?

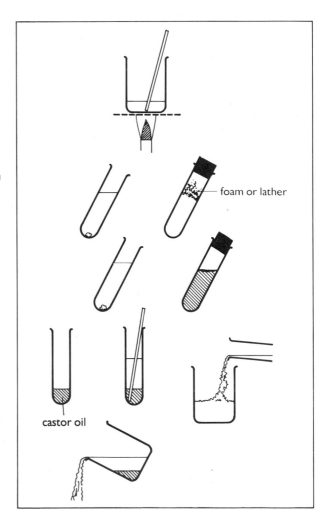

214

On page 94 we added soap to water. It gave a lather.

Distilled water needed very little soap. Lather formed at once. The water remained clear.

The tap water is hard. It contains dissolved calcium compounds. The soap we add reacts with them. Lather forms only when this reaction is over. So much more soap is needed. The water is cloudy with scum.

Detergents do not react with calcium compounds. Tap water and distilled water lather at once. Both need very little detergent. Both stay clear.

Lather produced by soap in hard and soft water.

The solid from castor oil behaves like soap. The oil we made behaves like a detergent. How does this happen?

oil + sodium hydroxide $\rightarrow$ soap
oil + sulphuric acid $\rightarrow$ detergent

We know that:
    ester + water $\rightarrow$ acid + alcohol
Sodium hydroxide speeds up the change. It also reacts with the acid. Its sodium salt is formed.

$$\begin{array}{c} H\ H\ H\ H\ H\ H\ H\ H\ H\ H\quad\quad O \\ | \ | \ | \ | \ | \ | \ | \ | \ | \ | \quad\quad\quad \| \\ H-C-C-C-C-C-C-C-C-C-C-C...C \\ | \ | \ | \ | \ | \ | \ | \ | \ | \ | \quad\quad\quad\quad \backslash \\ H\ H\ H\ H\ H\ H\ H\ H\ H\ H\quad\quad ONa \end{array}$$

The molecule has a long carbon atom chain. It is soap. We can call it sodium stearate. The alcohol formed with it is glycerol.

calcium +	sodium $\rightarrow$	calcium +	sodium
compounds	stearate	stearate	compounds
(hard water)	(soap)	(scum)	

$$\begin{array}{c} H\ H\ H\ H\ H\ H\ H\ H \\ | \ | \ | \ | \ | \ | \ | \ | \\ H-C-C-C-C-C-C-C-C....SO_3H \\ | \ | \ | \ | \ | \ | \ | \ | \\ H\ H\ H\ H\ H\ H\ H\ H \end{array}$$

A detergent molecule has the same shape. The group at the end of the chain comes from sulphuric acid. It does not react with calcium compounds. It does not form scum. It can lather at once in hard water.

Our skins produce an oily substance. This collects dirt. Oily dirt rubs off on clothes. Oil and water do not mix. Water alone does not wash off oily dirt. Soap and detergent help oil and water to mix.

Their long carbon chains are attracted to oil. The other end of the molecule goes into the water. The boundary between oil and water is broken up. The two mix. Rinsing removes both and also the dirt.

Vegetable oils can be used to make food. Turning them into detergent is a waste so oils from petroleum are used instead.

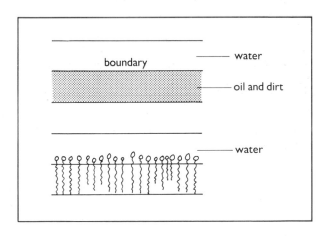

Cosmetics

The word cosmetic comes from a Greek word. It means 'to make beautiful'. Anything we use to make us more attractive is a cosmetic. Cosmetics have been used through the ages. Do you remember malachite?

It is the green ore copper carbonate. It was used as eye-shadow in Egypt at least 6000 years ago.

Do you wash?

Our skins produce a fatty substance called sebum. We also perspire. We wash these substances off. If we don't bacteria slowly decompose them. This gives rise to body odour. We don't think of soap as a cosmetic but shampoo is a cosmetic.

Our scalps produce sebum. Sebum collects dirt. Greasy dirt makes hair look dull and matted. Shampoo contains a good detergent to remove it. Colour and perfume make shampoo more attractive. Other extras are lemon juice, beer, egg – the list is endless. They don't help in washing hair but may have other useful effects.

Bath salts soften water. The crystals are sodium carbonate or sesquicarbonate. A coloured dye and perfume are usually added to the crystals. We use talcum powder to absorb water. It also makes our skin smooth or 'slippery'. It is mainly French chalk. Boric acid and fine chalk are also mixed in.

Do you clean your teeth?

A sticky layer called plaque forms on teeth. It harbours bacteria. Bacteria decompose the sugars in food. Acids are formed. They attack the enamel layer of teeth. This causes tooth decay.

Cleaning prevents decay in three ways. First, it removes food from between teeth. No food sugars will form on them. Secondly, it rubs off the plaque. This reduces the bacteria which use the sugar. Thirdly, it washes off acid already formed. Decay is cut down to a minimum.

Using toothpaste helps both effects. It must be soft but not gritty. Grit may damage the enamel. Very fine chalk is usually used. It neutralizes any acid already formed. This also helps to avoid decay. Water and glycerol keep the paste soft and moist.

A little dental soap in the paste helps cleaning. A flavour such as peppermint helps the taste. Some pastes contain fluorides. They reduce decay quite a lot. People still argue about their usefulness.

Going out into the world?

Setting lotion holds hair in place. So does hair lotion. Lotion contains a gum in solution. The solvent is mainly water. It contains some ethanol and glycerol. The ethanol is industrial meths.

Hair cream is mostly water and medicinal paraffin. A substance is added to make the oil and water mix. The mixture is an emulsion. Wax is dissolved in it.

Make-up includes lipstick, eye-shadow and mascara. It can make us more attractive. It may alter our appearance entirely. Make-up for stage and television often does this. We can 'put on a new face'.

Lipstick, eye-shadow and mascara provide colour. To do this well they must spread evenly and smoothly. They should also last and not easily rub off. All three contain an oil and a wax. This is to ensure even spread.

Lipstick contains beeswax in medicinal paraffin. It also has petroleum jelly. Colour and perfume are added. The melted mixture is poured into a mould. It cools and sets to a solid. This is a 'lipstick'.

Eye-shadow has a colour pigment. Mascara contains carbon black. Both are added to an oil and wax mixture. Thorough mixing is also important.

Finger and toe nails are often uneven. Nail varnish covers them with a smooth, hard layer. The varnish is a solution of shellac. This resin is dissolved in a mixture of liquids. A colour pigment is added. It is painted on the nail. The liquids evaporate quickly. This leaves a hard, coloured shellac layer. Meths and propanol are common solvents. They can be used to remove the shellac layer from the nail.

You can make these and other cosmetics. Two books will help. They are:

Cosmetic Chemistry by Anne Young (Mills and Boon)

Cosmetic Science by John Gostelow (Griffin and George).

Building large molecules

We have broken down large molecules into smaller ones. The reverse can also happen. Small molecules can be built into larger ones. We already know two ways of doing this.

1. Ethene is unsaturated. It reacts by **addition.** Its molecules join to those of other substances. The result is one single bigger molecule.

2. Alcohol and acid have active $-OH$ groups. These react by losing a molecule of water. The rest of the two molecules join up. An ester is formed. Its molecule is bigger than tjose of acid or alcohol. This is a **condensation** reaction.

Neither molecule shown is really large. The methods are important. They can be used to make huge molecules.

Polymers and plastics

Molecules of the same substance can react by method 1. Two ethene molecules link up. Their double bonds become single. Other C_2H_4 molecules join at the ends.

Thousands of molecules link up by addition. The result is a giant molecule. This must be a solid.

The prefix for one is mono–. The compound with single molecules is called a **monomer.** The prefix for many is poly–. The result of joining many single molecules is a **polymer.** The process is **polymerization.**

Ethene, C_2H_4, is the monomer. The polymer is poly–ethene. This has been shortened to polythene.

    ethene — addition polymerization → polythene
    (monomer)   (polymer)

Molecules of different compounds can be used. They also react by addition. The result is a co-polymer. Many polymers are plastics (see page 108).

Look at these formulae.

vinyl chloride   styrene   acrylonitrile

218

The names given are the older ones.
Are these four substances saturated?
What part of each formula gives the answer?
Will they react by addition?
Which ones are likely to polymerize by addition?

All four molecules contain a double bond. It makes them unsaturated. All four react by addition. All four can act as monomers. This is why the older names are given. *Polyvinyl chloride* is known as PVC.

chloroethene

$$\underset{H}{\overset{H}{\diagdown}}C=C\underset{Cl}{\overset{H}{\diagup}} \quad \underset{H}{\overset{H}{\diagdown}}C=C\underset{Cl}{\overset{H}{\diagup}} \rightarrow \quad \underset{H\ Cl\ H\ Cl}{\overset{H\ H\ H\ H}{-C-C-C-C-}}$$

These pipes are made from PVC.

Polymerization of styrene gives polystyrene. It is a very strong solid. It is light in weight. It does not conduct heat or an electric current. These properties give it many uses. Find out what they are.

$$\underset{H}{\overset{H}{\diagdown}}C=C\underset{C_6H_5}{\overset{H}{\diagup}} \quad \text{is phenylethene}$$

The polymer from acrilonitrile gives man-made fibres. Their trade names are well known: Orlon, Acrilan etc. Look for them on the labels on your clothes.

propenenitrile $\quad \underset{H}{\overset{H}{\diagdown}}C=C\underset{CN}{\overset{H}{\diagup}}$

The fluoride monomer is tetrafluorethene. It gives a polymer called Teflon or Fluon. It is very hard. It has few reactions. It is used as a non-stick layer, for example in saucepans.

$$\underset{F}{\overset{F}{\diagdown}}C=C\underset{F}{\overset{F}{\diagup}} \quad CF_2=CF_2$$

The polymers are named from the older names. The modern names are given with the formula. Perspex is a well known polymer.

Experiment 14.7 Polymer to monomer and back

Put perspex chips into a test tube to a depth of 3 cm. Fit the tube with stopper and delivery tube. Stand another tube in cold water. Put the delivery tube into it. Do the experiment in a fume cupboard.

Heat the perspex. Collect the liquid in the cooled tube. Stand the tube in hot water. Drop in a little di(dodecanoyl) peroxide. Let the mixture stand.

Which has the larger molecules, perspex or liquid?
What effect does the peroxide have on the liquid?

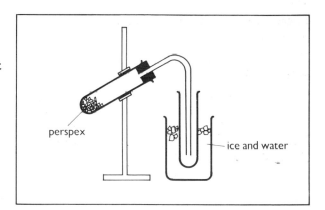

Perspex is an important polymer. It is a kind of super glass. It is used in making lenses. They are used in spectacles and optical instruments. The liquid has the smaller molecules. It is the monomer. The peroxide makes it polymerize to perspex again.

Polymers such as polythene are **plastics**. They can be moulded into almost any shape. They can be tailor-made for almost any purpose.

The second method

Method 1 uses substances with double bonds.
Method 2 uses compounds with active groups.
To show how it works we shall make nylon.
We need two different substances.

One is hexanedioyl dichloride. It is dissolved in tetrachloromethane. Call this solution A.

The other is hexane—1,6—diamine. It is dissolved in water. Call this solution B. A and B do not mix.

A dam wall made from plastic.

Experiment 14.8 To make nylon

Put 2 cm³ of A into a 5 cm³ beaker. Using a dropper add an equal volume of B. Add it carefully. It then forms a separate layer resting on solution A.

Gently put a pair of tweezers in the top layer. Pick up a bit of the boundary between A and B. Pull it slowly out of the liquid. Wind the thread round a glass rod. Slowly rotate the rod.

> A and B react. Where does the reaction happen?
> Describe the appearance of the thread formed.
> It is a solid. Are its molecules likely to be large?

A and B do not mix. They can react only where they meet. The tweezers pick up one product of this reaction. It is a solid. This means it may have large molecules. It is like a thread of white cotton. It is nylon thread.

For easy reaction we used a dichloride. It is made from an acid. In the manufacture of nylon the acid itself is used. Its molecule has an acid group at each end. Both ends of the molecules will react.

Diamine means that a molecule has two amine groups. This group is $-NH_2$. A nitrogen atom forms three covalent bonds. B has $-NH_2$ at each end of its molecule. Hexane means that a chain of six carbon atoms comes between them. Both ends react.

The acid and amine groups react. A molecule of water is formed. Free bonds are left at the ends of A and B. They join together. This is condensation.

This happens at both ends of B molecules. Both ends of A can react too. The product is a polymer.

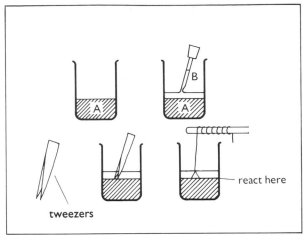

220

The new molecule is much larger. It still has an active group at each end. These groups react. The molecule becomes larger still. It still has an active group at each end. Two more units add on.

Very long chain molecules form. Each reaction adds another unit to the chain. It may be 100 units long. Amine and acid molecules alternate. Each has a chain of six carbon atoms. The product is nylon. Nylon is very tough and strong. It melts at about 250°C.

In this nylon each unit has six carbon atoms. It is called nylon 66. Other acids and amines can be used. The polymers formed will have slightly different properties. They will also be called nylons.

$$
\begin{array}{c}
\text{O} \quad\text{H H H}\ |\text{O}|\ \text{H H H H H H} \qquad \text{H} \\
\backslash\quad |\ |\ |\ |\ |\ |\ |\ |\ |\ |\ \diagup \\
\text{C-C-C-C-C}|\text{C-N}|\text{C-C-C-C-C-N} \\
\diagup\quad |\ |\ |\ |\quad |\ |\ |\ |\ |\ |\quad \backslash \\
\text{H-O}\quad\text{H H H}\quad|\ |\quad\text{H H H H H H}\qquad\text{H}
\end{array}
$$

the join

.... join-$(CH_2)_4$-join-$(CH_2)_6$-join-$(CH_2)_4$...

-A-join-B-join-A-

.... A-B-A-B-A-B-....

We used esters to explain condensation reactions. Suppose we use an alcohol with two −OH groups. Choose one with −OH groups separated by a carbon chain. Let it react with an acid at both ends.

Suppose we use an acid with two acid groups. Both ends of the molecule will be active. Both will react with the alcohol. Long chain molecules will form.

Terylene is an example of this kind of polymer.

Polyester coming off the machine.

One monomer is
$$
\begin{array}{c}
\text{H H}\\
|\ |\\
\text{H-O-C-C-O-H}\\
|\ |\\
\text{H H}
\end{array}
$$
Write it   HO⎯☐⎯OH

The other is $HOOC.C_6H_4.COOH$. Write it   HOOC⎯☐⎯COOH. Condensation gives an ester. This has an active group at each end of its molecule. These react by condensation forming esters.

HOOC ☐ CO |OH H| O-☐- O |H HO| OC-☐ CO |OH H| O-☐ .....

HOOC ☐ COO ☐ OOC-☐-COO-☐ -OOC- ....

A spinneret is used in the manufacture of terylene.

These polymers are called polyesters. The most well known example is terylene. It is very strong. Like nylon, it is a man-made fibre. Look for it on the labels on clothes.

Natural polymers

Many natural materials are polymers. For example, the sap from a rubber tree is a polymer. It is called latex. Acid turns it into solid rubber. The section on plastics will tell you more about it.

Starch is a polymer. Plants make it in their leaves. Potatoes and grain contain starch. Starch is part of our normal diet. It is insoluble. A soluble form can be made. We shall use it to test properties of starch.

Experiment 14.9 The properties of starch

Stir a pinch of soluble starch in a few cm³ of water. Pour this into 100 cm³ of boiling water. Put an iodine crystal into a test tube half full of water. Add a potassium iodide crystal. Shake the mixture. Use these solutions in the tests below.

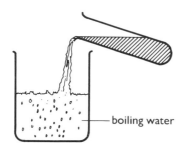

boiling water

a) **Add a drop of iodine solution to some starch solution.**
Put starch solution in a test tube. Add an equal volume of Fehling's solution. Boil it.

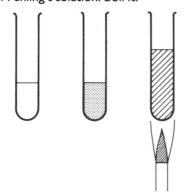

Which of these can be used as a test for starch?

Iodine and starch give a blue colour. It is used as a test for either of them. Starch does not affect Fehling's solution.

b) **Put 20 cm³ of starch solution into a beaker. Add 3 cm³ of dilute hydrochloric acid. Gently boil the solution. Remove a little of it every two minutes.**

Divide each part removed into two. Test one half with iodine solution. Boil the other with Fehling's solution.

c) **Fill a test tube one third full with starch solution. Dribble saliva into it. Shake the mixture. Take out a little every two minutes. Test each part as in b). Chew a small piece of bread for several minutes.**

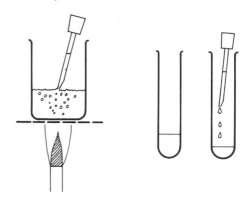

Is starch used up in the acid solution?
Does saliva cause the starch to disappear?
Does the taste of the bread change?

With each sample, iodine gives a paler blue than the sample before. Starch is being used up. Acid and saliva turn it into sugars. This is why the taste of bread becomes sweeter. Saliva contains an enzyme, ptyalin. It is a catalyst for the change starch → sugars. This change is part of digesting our food.

Some sugars react with Fehling's solution. The blue colour of copper(II) ions goes. An orange solid, copper(I) oxide, is formed.

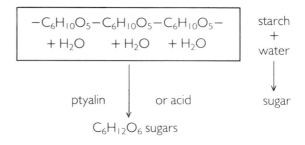

$$-C_6H_{10}O_5-C_6H_{10}O_5-C_6H_{10}O_5-$$
$$+H_2O \qquad +H_2O \quad +H_2O$$

starch + water

ptyalin   or acid   sugar

$C_6H_{12}O_6$ sugars

Things to do

Make a list of all plastic articles in your home. Find out the names of the polymers they contain. Check the labels on your clothes. What fibre does each of them contain? Do they contain wool or cotton?

Summary

Large molecules can be broken down into smaller ones. Using a catalyst, ethanol decomposes. It gives a gas called ethene. This is a hydrocarbon, formula C_2H_4. Ethene burns with a yellow flame.

Its molecule has a double bond. It is unsaturated. It reacts by addition. In addition reactions, two or more compounds form only one product.

$$C_2H_4 + Br_2 \rightarrow C_2H_4Br_2$$

Ethene molecules react with each other. They form long chain molecules. The product is a solid. It can be moulded into shape. It is called polythene.

The process is polymerization. It makes large molecules from small ones. The 'small molecule' compound is a monomer. The product is a polymer. Substances like ethene polymerize by addition.

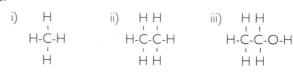

Condensation is also used. In this, two different molecules have active groups. These groups react. A small molecule like H_2O is formed. The rest of the molecules link up. This gives a big molecule.

Two monomers are used. Their molecules have two active groups, one at each end. These groups react. Chain molecules form. This is the polymer.

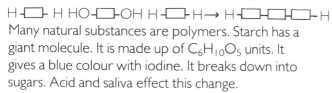

Many natural substances are polymers. Starch has a giant molecule. It is made up of $C_6H_{10}O_5$ units. It gives a blue colour with iodine. It breaks down into sugars. Acid and saliva effect this change.

Questions

1. Ethene is an unsaturated compound.
   a) What does 'unsaturated' mean?
   b) Draw a simple structural formula for ethene.
   c) Ethene reacts with hydrogen to form a saturated compound, ethane. What does 'saturated' mean?
   d) Write an equation for the last reaction.
   e) Draw a simple structural formula for ethane.
   f) Why are ethane and ethene called hydrocarbons?

2. Ethene is an unsaturated hydrocarbon.
   a) What does the term 'unsaturated' mean?
   b) What does the term 'hydrocarbon' mean?
   c) Bromine is added to ethene. What would you see?
   d) Ethene can be polymerized. What is a polymer?
   e) Name the polymer formed from ethene.
   f) Why is this reaction called 'addition' polymerization?

3.

i)	ii)	iii)
H H-C-H H	H H H-C-C-H H H	H H H-C-C-O-H H H

iv)	v)	vi)
H C=C H H H	H H H-C-C-H Br Br	H O H-C-C H O-H

   a) Name the compounds from i) to vi).
   b) Three of them are hydrocarbons. Which ones?
   c) One of these hydrocarbons is unsaturated. Which?
   d) What everyday substance is made from substance iv)?
   e) Which two substances react to form an ester?
   f) Substances ii) and iv) both burn in air. Name the products formed by burning.
   g) Apart from burning the reactions of ii) and iv) are different. Explain why.
   h) Which of the six substances is made by fermentation? What part do enzymes take in fermentation?

223

The plastics age

What time did you get out of bed?
What plastics did you use in the next hour?
Try to work out how many you use in a day.

A plastic is a substance which can be moulded into shape. Most plastics are polymers. They are manufactured in huge amounts. Many different monomers are used. There are many methods of manufacture.

Two types of reaction are used. Addition reactions use molecules with double bonds. They may be of one substance only. Two monomers may also be used together. In condensation two monomers are used. Each has active groups. These react. Both methods build long chain molecules. Polymers have giant molecules.

Their properties depend on three main factors: a) which monomers are used, b) the method used to make them react, c) cross linking between the chain molecules. So a plastic can be tailor-made. The factors can be chosen to give the properties needed.

Take polythene as an example. Solvent is fed into a reaction vessel. A catalyst is added. Ethene under pressure is passed in. The temperature used is about 75 °C. Ethene reacts to form the polymer.

It leaves the reaction vessel dissolved in the solvent. Any catalyst is removed. The solvent is evaporated. The solid plastic is formed. It is made into small pieces or granules. These can be heated.

Heating gives the chain molecules greater movement. The plastic softens. It can be shaped by a mould. On cooling the plastic hardens in this shape. It can also be made into thin film or sheet.

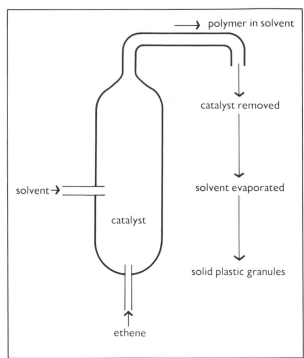

Polythene is a thermoplastic. Heat softens it. It can be moulded. It can be heated, softened and remoulded again. For some plastics this is not true. They are called thermosetting. Heat softens them. They can be moulded. The moulded substance will not soften on heating. The chains have become cross linked. They are held in rigid positions.

Plastics are moulded in many ways. One is called extrusion. Heated plastic is forced through a gap. Pipes and tubing are made in this way. So are rods, plastic sheet, tape and plastic coated wire.

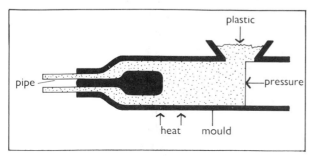

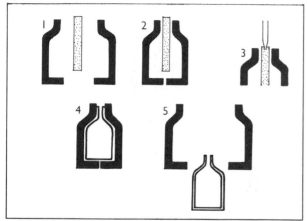

Blow moulding is used to make hollow objects.

1. Heated plastic is fed into a mould.
2. The mould closes.
3. An air inlet is put into the plastic.
4. Air is blown in. The plastic takes the shape of the mould.
5. When the mould opens it drops out.

Rubber is a polymer made from latex. Some of this comes from rubber trees. Most of it is now man-made. Monomers such as butadiene are used. They are polymerized with monomers such as styrene. The polymer formed still contains double bonds.

$$H\ H\ H\ H \qquad H\ H\ H\ H\ H\ H\ H\ H$$
$$C{=}C{-}C{=}C \rightarrow -C{-}C{=}C{-}C{-}C{-}C{=}C{-}C-$$
$$H \qquad H \qquad H \qquad H\ H \qquad H$$

bewt-a-di-een

All rubbers are elastic. Forces applied to them change their shape. The forces do this by pulling the chain molecules apart. The molecules return to where they were when the forces are taken away. Cross linking between chains improves this effect.

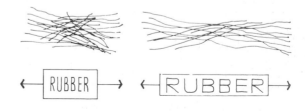

Sulphur atoms can link carbon atoms in different chains. Carbon atoms link directly as well. Both make the rubber more elastic. It improves strength and toughness. Cross linking makes it less plastic.

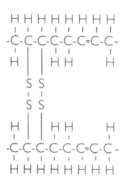

Rubber, sulphur and carbon black are mixed. The mixture is fed into a heated mould. It forms a hard, tough elastic shape. Tyres are made in this way. The process is called vulcanization. Foam rubber needs no mould. A substance is added to produce gas. The heated mixture forms a froth. It cools to a solid.

Plastics have thousands of uses. They are needed for contact lenses, tennis rackets, nylon rope, nylon stockings, pop records, T–shirts, roof insulation, floor carpets, tape decks, trousers and many more.

Drugs and you

Read the section on Chemistry and disease again.

Many substances have medical uses. They help to keep the body healthy. They deal with physical illness. We group them loosely together as 'drugs'.

Some supply missing substances. Others destroy disease organisms. A third type relieves disease symptoms. Pain killers belong to this group.

People suffer from mental illness. Drugs are used in mental healing. They affect the body as well.

What is a drug?

We are alive. Every bit of us has some use. Each bit has a job, or 'function'. Some have more than one. Taking in a substance may affect a function. If it does, the substance is a drug.

A drug modifies some function of mind or body.

So alcohol is a drug (page 206). Smoking tobacco acts as a drug. They affect mind or body, or both.

Where do drugs come from?

Some come from plants and animals. Some are made from minerals. Most of them are manufactured. Natural ones have been known for centuries. One of these is cannabis. It is made by a plant, Indian hemp. It is extracted from the leaves and flowers.

How are they used?

As medicines we use them in small doses. We swallow them as tablets or pills. Others are liquids or solutions. We drink them or have them injected.

Some, such as aspirin, are freely on sale. Others need a doctor's prescription. Use both exactly as you are told. Never take more than the label says. Even aspirin can do damage. Large amounts can kill.

Are there other uses?

Healthy people don't need drugs. But some take them just to try. They do this for many reasons. Some are curious. They wonder what will happen. Some are bored. They want something new to happen. Some are unhappy. They think they might enjoy it. Most first-timers don't like to refuse.

People smoke for the same reasons. Someone offers a first cigarette. Someone offers a first fix too. Don't accept either. The risk is too big.

How do they work?

They can be swallowed, smoked, sniffed or injected. The drug gets into the bloodstream. It is carried to the brain. The brain is affected by it. Nerves connect brain and parts of the body. Messages pass along them. The functions of both are affected.

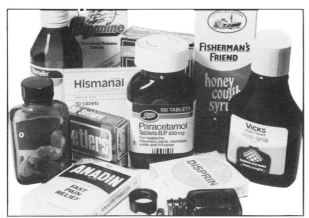

Some useful drugs we can buy to treat ourselves.

Smoking cigarettes is harmful and unpleasant. It also affects people around us.

What happens?

Four factors decide this:

1. the type of drug,
2. the amount taken,
3. how the person feels at the time,
4. his or her situation at the time.

Use alcohol as an example. Taking small amounts can be good. Happy people become happier. Friends are even more friendly. But angry people become furious. Anger can lead to dangerous violence.

Large amounts have a greater effect. People become drunk. They stagger. Their speech is blurred. They react more slowly. The mind has lost control of the body. Concentration has gone. A drunk is too fuddled to know this. This is when accidents happen. Most drugs reduce concentration and slow reactions down.

Long term effects can be tragic. People can't do without alcohol. They become 'addicted'. They are 'alcoholics'. The cure is possible but very hard. Brain, liver and kidney can be damaged. Using one drug may lead to using others. Drugs destroy people.

Bad use of drugs is called 'drug abuse'. Some common examples are shown in the table. CR means concentration reduced. RS means reaction slowed down. How long the effects last is shown by 'Hours' and 'Minutes'.

Name of drug	How taken	Effects of using once	Effect of using regularly	Against the law or not
Cannabis, pot, hash, marijuana, grass, dope.	Smoked with or without tobacco.	CR. RS. Makes things brighter and louder. Hours.	Possible lung cancer. Drug builds up in the body.	Illegal to have or use it.
Opiates, heroin, opium.	Swallowed, injected, smoked or sniffed.	CR. RS. Can kill you especially with alcohol. Hours.	Vein and skin damage. Easy to become addicted. Mental problems. Cure very hard.	Illegal without doctor's prescription. Many addicts die.
Amphetamine, pep pills.	Swallowed, sniffed or injected.	Alert, lively. Tired later. A large dose can kill. Hours.	Easy to become addicted. Hard to give up.	Illegal without doctor's prescription.
Solvent sniffing.	Vapour sniffed by nose.	CR. RS. Makes drunk. Can kill at first try. Minutes.	Vomiting. Mouth sores. Blurred vision. Death?	Legal. Results can lead to police arrest.

Nitrogen

Write down all you know about the element nitrogen.

Experiment 15.1 The properties of nitrogen

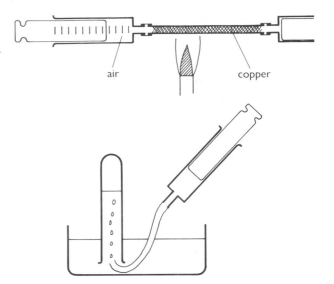

Fill a tube with tiny pieces of copper wire. Connect a syringe at each end. Have one syringe full of air. Read off its volume. Heat a section of the copper. Push the air through it.

Heat the next section. Push the air back into the first syringe. Do this until the heated copper shows no change. Read off the volume of air remaining. Bubble it into test tubes.

Test the gas with a wet litmus paper, a lighted splint and burning magnesium. Invert a test tube of gas in dilute acid. Invert a second test tube of gas in sodium hydroxide solution.

   What gas remains in the syringe?
   What percentage of air is this gas?
   Is it a reactive element?

Air is mainly oxygen and nitrogen. Heated copper takes out the oxygen. About 78% of air is nitrogen. 0.03% is carbon dioxide. Carbon dioxide is in the nitrogen we tested. It is too small to matter.

Only magnesium reacts with nitrogen. It almost goes on burning. It combines to form a white solid. This is magnesium nitride. Nitrogen puts out a lighted splint. It does not burn. It does not affect litmus. It does not react with acid or alkali. It is not a reactive element.

The atomic number of nitrogen is 7.
Its relative atomic mass is 14. Its position is shown.

Element	C	Nitrogen	O	F	Neon	Na	Mg	Al	Si
Atomic Number	6	7	8	9	10	11	12	13	14

Answer these questions:
   What is the symbol for nitrogen?
   What do atomic number and atomic mass mean?
   How many protons does a nitrogen atom contain?
   How many neutrons are there in its nucleus?
   How many electrons does a nitrogen atom have?
   How many electrons fill the first shell?
   How many does this leave for the second shell?
   What is the electron structure of nitrogen?
   What is the electron stucture of the nearest noble gas?

Atomic Number = number of protons in the atom.
Number of electrons = number of protons.
Number of neutrons = atomic mass − atomic number.

the N atom has 7 protons
7 electrons
14 − 7 = 7 neutrons

an outside shell of five electrons

The electron structure of a nitrogen atom is 2.5.
For the nearest noble gas, neon, it is 2.8.
A nitrogen atom will gain three electrons.

The structure of the magnesium atom is 2.8.2.
What is the formula of magnesium nitride? It is a solid. What type of
bond is likely to be holding its atoms together?

A solid is likely to be ionic. One magnesium atom
loses two electrons. It becomes the ion Mg^{2+}. A
nitrogen atom can gain three. It becomes N^{3-}. How
can gain and loss be made equal? Use three Mg
atoms. They supply 6 electrons. These satisfy two N
atoms. $3Mg^{2+}$ join $2N^{3-}$

$$Mg\,(2.8.2) \rightarrow Mg^{2+} + 2e$$
$$N\,(2.5) + 3e \rightarrow N^{3-}$$

$$3Mg^{2+} . 2N^{3-} \quad \text{or} \quad Mg_3N_2$$

The change from N to N^{3-} needs a great deal of
energy. Only very active metals can supply it. They
give energy when they turn into ions. Calcium,
sodium and magnesium burn in nitrogen.

Nitrogen and food

Experiment 15.2 Heating foods with soda lime

Heat a little powdered meat in a small test tube. Smell
with care any gas given off. Hold a wet litmus paper in
the mouth of the tube. Test any gas with a lighted splint.

Mix a little powdered meat with soda lime. Heat this in
the same way. Use the same tests. Use other foods in
place of meat. Milk, corn flakes, sugar, crisps, bread and
fish may be used. They should be dry and powdered.

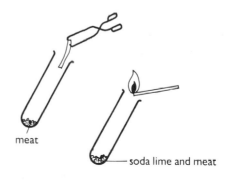

meat

soda lime and meat

> The smell suggests that a gas is given off.
> Do you recognize it by its smell?
> Is it given off by all foods or only some?
> What happens to the litmus and the lighted splint?
> Are other substances formed during the heating?

Foods contain carbon compounds. This is why heating
makes then char. Condensation forms in the tube. This is
probably water. Some foods give a gas on heating. It has
a pungent smell. It turns litmus blue. It must be the gas
ammonia. Ammonia has the same smell as smelling salts.

Not all foods give ammonia. Only those which contain
protein do so. Soda lime makes the gas come off more
quickly.

Ammonia

We can buy ammonia as a concentrated solution in water. The solution is called aqueous ammonia.

Experiment 15.3 To find out what elements ammonia contains

Put aqueous ammonia in the tube with a side neck. Add bits of pumice. Gently heat it to get ammonia. Pass the gas through quicklime to dry it.

At the same time heat the black copper oxide. Collect liquid in the cooled tube. Collect gas over water.

Test the boiling point of the liquid. Try other tests on it. Test the gas with litmus. Find out if magnesium burns in it.

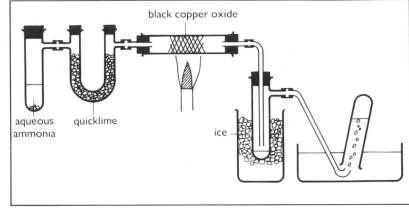

black copper oxide

aqueous ammonia   quicklime   ice

What happens to the heated copper(II) oxide?
What substance is the oxide giving up?
Does its boiling point show what the liquid is?
What other tests did you use on the liquid?
What is the gas collected in the test tubes?
Complete the word equation
ammonia + copper(II) oxide → ? + ? + ?

The gas puts out a lighted splint. Magnesium nearly burns in it. Litmus is not changed. The gas is nitrogen.

The boiling point of the liquid is 100°C. It could be water. It turns white anhydrous copper sulphate blue.

The copper oxide is reduced to copper. It gives up oxygen.

Where does the water come from? The ammonia is dried. Water must be a product of the reaction. Its oxygen comes from the copper(II) oxide. Its hydrogen must come from ammonia. The reaction is

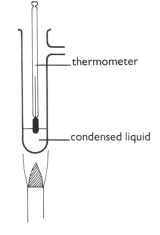

thermometer

condensed liquid

$$\begin{array}{ccccccccc}
\text{ammonia} & + & \text{copper oxide} & \rightarrow & \text{nitrogen} & + & \text{water} & + & \text{copper} \\
? & + & CuO & \rightarrow & N_2 & + & H_2O & + & Cu
\end{array}$$

Ammonia is a compound of nitrogen and hydrogen. You know their electron structures. Use them to work out a formula for ammonia.

The electron structure of nitrogen is 2.5. Its atom therefore needs three electrons to reach 2.8. It can get them by sharing with other atoms. Three shared pairs make three covalent bonds. The formula of the gas ammonia must be NH_3.

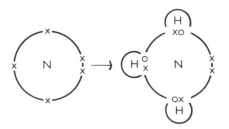

*Experiment 15.4 To make and test ammonia

Mix equal amounts of slaked lime and ammonium chloride. Heat the mixture in a round bottomed flask. Dry the gas with quicklime in a tower. Fill four gas jars with the gas. Use a damp litmus to find out when each jar is full.

Invert a jar of gas in water. Remove the cover. Test the water with an indicator.

Invert a gas jar in dilute acid. Remove the cover.

Make a gas jar of hydrogen chloride. Invert it over a gas jar of ammonia. Remove the covers.

Hold a lighted splint over a gas jar of ammonia. Remove the cover from the jar. The ammonia rises.

> Does water rise into the gas jar of ammonia?
> Does the dilute acid rise into the jar?
> How well does the gas dissolve in each case?
> What do ammonia and hydrogen chloride form when mixed?
> What happens to the lighted splint?
> Did you see anything else?

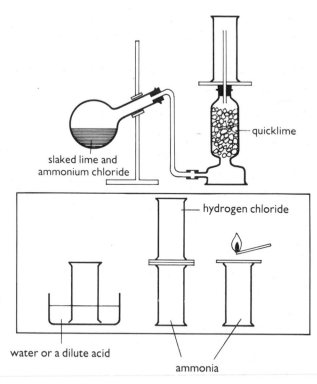

slaked lime and ammonium chloride

quicklime

hydrogen chloride

water or a dilute acid

ammonia

Fit a round bottomed flask as shown. Close the clip on the rubber tubing. Take out the stopper. Fill the flask with ammonia. Dip the end of the glass tube In water. Stopper the flask with it. Colour water with Universal Indicator. Put the rubber tube into the water. Open the clip.

> What do you see?
> Why does it happen?

Water rises into the gas jar. So does dilute acid. Ammonia must dissolve (and react?) in each liquid.

Water rushes into the flask. A fountain is formed. This shows that almost all the ammonia has gone. It can only have dissolved in water from the tube. It must be very soluble. The solution is an alkali.

Ammonia and hydrogen chloride give white smoke. This settles on the jar as crystals like salt. Ammonia puts out a lighted splint. A pale yellow flame lingers above the splint. Ammonia nearly burns in air.

Ammonium compounds

Here is the ammonia molecule again. Its atoms are joined by covalent bonds. Each bond is a shared pair of electrons. Both atoms supply one electron of the pair.

nitrogen N   ammonia NH_3

An ammonia molecule has an unused pair of electrons. This spare pair can be used for bonding. For example, a hydrogen ion, H^+, has no electrons. It can bond to ammonia using the spare pair. The result is NH_4^+. It behaves like the sodium ion, Na^+. It is therefore called the ammonium ion.

$$H \overset{x}{\underset{o}{\overset{o}{\underset{x}{N}}}} \overset{x}{x} H + H^+ \rightarrow \left[H \overset{x}{\underset{o}{\overset{o}{\underset{x}{N}}}} \overset{x}{x} H \right]^+$$

unused pair NH_4^+

The bond is still a shared pair of electrons. Both electrons are supplied by one of the atoms. This explains why aqueous ammonia is an alkali. An ammonia molecule takes an H^+ ion from water. This leaves a hydroxyl ion, OH^-. Ammonia puts OH^- ions into water.

$$NH_3 + H_2O \rightarrow NH_4^+ + OH^-$$

$$H \overset{x}{\underset{o}{\overset{o}{\underset{x}{N}}}} \overset{x}{x} H + H_2O \rightarrow \left[H \overset{x}{\underset{o}{\overset{o}{\underset{x}{N}}}} \overset{x}{x} H \right]^+ + OH^-$$

$$NH_3 + H_2O \rightarrow NH_4^+ + OH^-$$

Ammonia and hydrogen chloride react. White crystals form. An ammonia molecule takes an H^+ from HCl. Together they form NH_4^+. This leaves a Cl^- ion. The ions attract each other. They form a crystal lattice, as in sodium chloride. The white crystals are a salt. It is called ammonium chloride.

$$H - \overset{x}{\underset{|}{N}} \overset{x}{x} \quad H \overset{x}{\underset{o}{\overset{x}{\underset{x}{Cl}}}} \overset{x}{x} \rightarrow NH_4^+ + Cl^-$$

$$NH_4^+ \ Cl^-$$

$$NH_4Cl$$

This bonding shows how ammonia reacts with acids. Nitric acid contains H^+ and NO_3^+ ions in solution. NH_3 reacts with H^+. The solution now contains NH_4^+ and NO_3^- ions. When it evaporates crystals form. They are the salt ammonium nitrate.

$$H - \overset{x}{\underset{|}{\overset{|}{N}}} \overset{x}{x} \quad + (H^+ \ NO_3^-) \rightarrow NH_4^+ \ NO_3^-$$

Experiment 15.5 Heating ammonium salts

Heat a few ammonium chloride crystals in a dry test tube.

Put some crystals in a second test tube. Stick a wet litmus paper just above them. Put in some rocksill fibre. Put a second litmus paper just above it. Heat the crystals. Watch the litmus.

What happens to the heated crystals?
What changes in the litmus papers did you see?
Can you work out what these changes mean?

*Put a small pinch of ammonium nitrate crystals in a dry test tube. Hold it in a clamp. Put safety screens around it. Heat it gently.

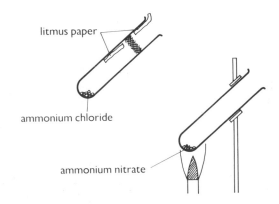

litmus paper

ammonium chloride

ammonium nitrate

What happens?
What may one of the products be?

Heat does not melt ammonium chloride. No liquid forms. The crystals slowly disappear. They form again on the sides of the tube.

The lower litmus turns blue. Ammonia must have caused this. The crystals must have decomposed. Then hydrogen chloride must be there as well.

Both move up the tube. Ammonia diffuses faster. Experiment 7.2 shows this. More of it gets to the litmus first. This is why both turn blue. More hydrogen chloride is left behind. The lower litmus turns red. Up the tube the gases cool. They re-combine. Ammonium chloride crystals form.

$$NH_4^+.Cl^- \xrightleftharpoons[\text{cooling}]{\text{heating}} NH_3 + HCl$$

The reaction is reversible.

Ammonium nitrate decomposes. It explodes gently. It may blow smoke rings. One product looks like steam. The other is a gas. It is called dinitrogen oxide. It was once used as an anaesthetic called laughing gas. Its formula is N_2O.

$$NH_4NO_3 \text{ or } (NH_4^+.NO_3^-) \rightarrow N_2O + 2H_2O$$

The uses of ammonia and its compounds

Ammonia has very many uses.
It is a grease remover. It also softens water. It is used in the manufacture of many substances. They include washing soda, urea and nitric acid. Plastics are made from monomers. Some of these are made from ammonia. A nylon monomer is one.

Ammonium salts come from ammonia. It is passed into an acid solution. The neutral solution is then evaporated. Crystals of the salt appear.

Ammonium chloride is used in dry cells. It is used as a flux. It cleans metal surfaces before soldering.

Ammonium nitrate is used as an explosive. It can be used alone or mixed with other substances. It is an important fertilizer. Mixed with chalk ammonium nitrate is called nitrochalk. Chalk makes acid soils neutral. Chalk is also there to reduce the risk of explosion.

Ammonium sulphate is a famous fertilizer. It is called 'sulphate of ammonia'. Ammonia and its salts put nitrogen into the soil. Large amounts of them are used. Read the section on fertilizers.

All these uses need large amounts of ammonia. A cheap method of making it on the large scale is needed. It is made from nitrogen and hydrogen.

Packing bags of fertilizer.

Sulphate of ammonia silo.

233

The manufacture of ammonia

Ammonia is important. About 8 million tonnes of it are manufactured in Britain every year. Nitrogen and hydrogen are used. They react very slowly. The reaction is reversible. It gives out heat. ΔH for an exothermic reaction is minus.

$$N_2 (g) + 3H_2 (g) \rightarrow 2NH_3 (g) \quad \Delta H = -92 \text{ kJ}$$

What methods can we use to speed up a reaction? Heating is one way. Another is to use a catalyst. We can combine both methods. The gases are passed through hot iron. This acts as the catalyst. It also makes the gas mixture hot.

Look at the equation.
$N_2 + 3H_2$: four molecules of the gases react.
$2NH_3$: only two molecules of ammonia form.
Volume is proportional to number of molecules. The ammonia volume drops to half. Pressure can affect this kind of change. High pressure makes it move to smaller volume. More ammonia forms.

Nitrogen is made from liquid air. Hydrogen is made from oil or natural gas. The two gases are mixed. They are compressed to 250 times air pressure. The mixture is fed into a strong steel cylinder. This contains the catalyst. The cylinder is at 500°C. It heats the gas mixture.

The gases react. A mixture leaves the cylinder. It contains about 12% ammonia. The rest is unused nitrogen and hydrogen. It cools down. Cooling under pressure turns ammonia into a liquid. The unused gases pass on. They are used again. Ammonia is stored under pressure in liquid form. This method of manufacture is the **Haber Process.**

Ammonia manufacture at Billingham.

Summary

78% of air is nitrogen. Its molecule is diatomic. It is written N_2. Nitrogen is not reactive. It puts out a lighted splint. It does not burn. Nitrogen does not react with acid, alkali or lime water.

Its electron structure is 2.5. When it reacts a nitrogen atom needs to gain three electrons. Active metals burn in it. The ion N^{3-} is formed.
$$3Mg + N_2 \rightarrow 3Mg^{2+}.2N^{3-}$$
$$(Mg_3N_2) \text{ magnesium nitride}$$

Protein foods heated with soda lime give the gas ammonia. It is also made by heating slaked lime with ammonium chloride. Ammonia turns moist litmus blue. It can be recognized by this and by its smell.

Ammonia reduces hot copper(II) oxide. Copper is formed. Water forms. Its hydrogen must have come from ammonia. A gas is collected. Tests prove it is nitrogen. It can only have come from ammonia. Ammonia is a nitrogen-hydrogen compound.

Ammonia has the formula NH_3. The atoms are joined by covalent bonds. NH_3 has a 'spare pair' of electrons. With them it bonds to a hydrogen ion. The result is an ammonium ion, NH_4^+.

$$\begin{matrix} & H & & & \\ H\text{-}N & \overset{\times}{\underset{\times}{}} & + & H^+ & \rightarrow & NH_4^+ \\ & H & & & \end{matrix}$$

Ammonia is very soluble in water. Aqueous ammonia is an alkali. It contains hydroxyl ions, OH^-
$$NH_3 + H_2O \rightarrow NH_4^+ + OH^-$$
Ammonia also reacts with acids. The products are ammonium salts.
$$NH_3 + HCl \rightarrow NH_4^+.Cl^-$$

Ammonia is manufactured by the Haber Process. Nitrogen and hydrogen are mixed and compressed. They pass through a catalyst, iron, at 500°C. As the products cool, ammonia turns to a liquid. It separates from unused gases which are used again. The reaction is exothermic.

reactants
catalyst
temperature
separation

$$N_2(g) + 3H_2(g) \rightarrow 2NH_3(g) \quad \Delta H \text{ is minus}$$

Ammonia is used to make other substances. It and its salts are fertilizers. They supply nitrogen to the soil. Ammonium salts decompose on heating.

Questions

1. Salt A consists of white crystals. It is heated with slaked lime. A colourless gas, B, is formed. B has a sharp smell. It turns litmus blue. It is very soluble in water giving solution C. Sulphuric acid is added to C. When the solution is neutral it is evaporated. White crystals of solid D remain.

   Name one salt A which gives these results.
   Name the gas B and give its formula.
   What ions are present in solution C?
   Name the white solid D.
   B is made on the large scale from two gases, E and F. Name these gases.
   They are passed through a catalyst. Name the catalyst. Explain what the word catalyst means.

2. Ammonium nitrate is a fertilizer. What element does it put into the soil?
   Why do plants need this element?
   Explain two natural ways in which this element gets into soil.
   Ammonium nitrate fertilizer is usually mixed with chalk. Give one reason why.

3. Ammonia is heated with copper(II) oxide. The oxide is reduced to copper. A colourless liquid is formed. It boils at 100°C. What element does this show to be present in ammonia? Give reasons.

4. In each part, five answers are given. They are labelled A, B, C, D and E. Only one of these answers is correct. Choose the correct answer.

   a) A certain gas puts out a lighted splint. It does not affect lime water. It does not burn. The gas is
   A carbon dioxide
   B hydrogen
   C nitrogen
   D oxygen
   E natural gas

   b) All these statements about ammonia are true, except one. Which statement is not true?
   A Ammonia turns moist litmus blue
   B Ammonia is very soluble in water
   C Ammonia has a recognizable smell
   D Ammonia reacts with hydrogen chloride
   E Ammonia is an ionic (electrovalent) compound

   c) Ammonia dissolves in water. The solution is an alkali because
   A Ammonia is a covalent compound
   B The solution contains hydrogen ions, H^+
   C The solution contains hydroxyl ions, OH^-
   D The solution smells of ammonia
   E It turns litmus red

   d) A food is heated with soda lime. It gives off ammonia. This kind of food is
   A a carbohydrate
   B a fat
   C a protein food
   D a mineral salt
   E sugar

5. These compounds are fertilizers :
   a) ammonia, NH_3. b) ammonium nitrate, NH_4NO_3.
   c) ammonium sulphate, $(NH_4)_2SO_4$.
   $H = 1; N = 14; O = 16; S = 32; Ca = 40$
   Calculate the formula mass of each compound. Work out the percentage of nitrogen in each. Give one other use of each.

6. Ammonia reacts with hot copper(II) oxide. Copper forms. Is this change reduction? A liquid forms. Give two tests to prove that it is water. A gas is collected. It puts out a lighted splint. It does not burn. It has no effect on lime water. Which gas is it? What does this tell us about ammonia?

Ammonia and nitric acid

Ammonia is less dense than air. It rises out of a gas jar. It puts out a lighted splint. It almost burns in the air above the jar. It may burn in oxygen.

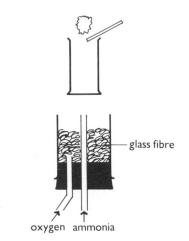

*Experiment 15.6 Ammonia and oxygen

Set up the apparatus shown in the diagram. Pass ammonia up the middle tube. Pass oxygen up the other tube. Hold a lighted splint above the middle tube. Reverse the gas connections. Hold a lighted splint above the middle tube again.

Are the two flames exactly the same?

The fibre spreads the oxygen. It surrounds the middle tube. Ammonia burns in it with a yellow flame. Oxygen burns in ammonia. The two flames are alike. This tells us what a flame is.

A flame is the space in which gases react. The reaction gives heat and light.

Ammonia does not burn in air. It does burn in pure oxygen. It may react with air if we use a catalyst.

Platinum gauze catalyst.

*Experiment 15.7 Ammonia and air with a catalyst

Take a flask. Put in a little concentrated aqueous ammonia. Hold a coil of platinum wire above the neck. Put a piece of white paper behind it.

Blow air into the flask. Briefly heat the coil. Touch it with a flame or pass current through it. Alter the conditions to get results. Increase the air stream. Dilute the ammonia solution. Hold a wet neutral litmus paper above the coil.

Did you see any new substance?
Did the white paper let you see its colour?
What happens to the indicator papers?
Is there any change in the platinum?
Is there any chemical change?
If so, is it exothermic or endothermic?

The solution gives ammonia gas. It mixes with the air. The mixture passes over the platinum. This acts as a catalyst. It glows dull red. The two gases must be in the right proportions.

The best results show a brown gas. The white paper makes it easier to see. It is nitrogen dioxide. It dissolves in water. Litmus shows it is acid.

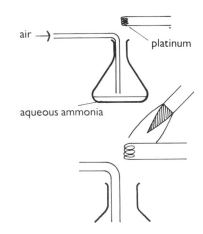

ammonia + oxygen → brown gas from air

brown gas + water → acid solution

Ammonia reacts with oxygen in the air. The reaction is exothermic. It keeps the coil hot. Too much ammonia may be present. It can react with nitrogen dioxide. No brown gas will be seen. This is why we altered the conditions.

In Experiment 15.6 ammonia burns in oxygen. Nitrogen and steam are formed.

$$4NH_3 + 3O_2 \rightarrow 2N_2 + 6H_2O$$

With a catalyst the products are different. The brown gas is nitrogen dioxide. It reacts with water to give an acid solution. It contains two acids.

One is nitric acid, HNO_3. The other is nitrous acid, HNO_2. Air slowly oxidizes it to nitric acid.

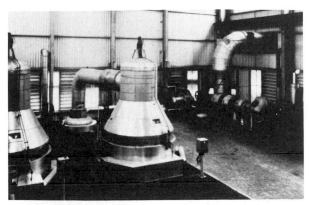

Manufacturing nitric acid.

The manufacture of nitric acid

The method of Experiment 15.7 is used. Ammonia is mixed with ten times its volume of air. Reaction happens in a converter. The catalyst is wire gauze. This is stretched across the converter.

The gauze is 90% platinum. It is heated electrically. Then heat from the reaction keeps it at about 900°C. Ammonia and air react to form nitrogen monoxide, NO.

$$4NH_3 + 5O_2 \rightarrow 6H_2O + 4NO \quad \Delta H \text{ is minus}$$

The gases which leave the converter are cooled. Nitrogen monoxide reacts with oxygen from air. The brown gas nitrogen dioxide, NO_2, is formed.

$$4NO(g) + 2O_2(g) \rightarrow 4NO_2(g)$$

It is mixed with more air and dissolved in water. Only nitric acid is formed in solution.

$$4NO_2(g) + 2H_2O(l) + O_2(g) \rightarrow 4HNO_3(aq)$$

In solution its ions are H^+ and NO_3^-

Nitric acid has many important uses. So have its salts. They are called nitrates. They all contain the nitrate ion, NO_3^-. They all dissolve in water.

Grace: I'm making nitric acid. Get the catalyst.
George: I got the cat a list. He can't read it though.

237

Nitric acid and nitrates

Sodium nitrate is found in nature. Large amounts once existed in Chile. It was used to manufacture nitric acid, before the ammonia method was used. It is also a well known fertilizer.

*Experiment 15.8 To make nitric acid

All the apparatus must be made of glass. Put a teaspoonful of sodium nitrate crystals in the flask. Cover them with concentrated sulphuric acid. Gently heat the mixture. Collect any liquid in a cooled test tube.

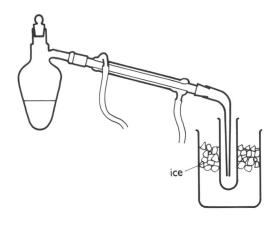

ice

Use it for three tests. Do these in a fume cupboard. Hold a white screen behind them.

a) Add a little of the liquid to copper in a test tube.
b) Heat some of the liquid in a crucible. When it boils drop in a red hot piece of charcoal.
c) Heat a small pile of sawdust on a tin lid. Add drops of the liquid to it using a dropper.

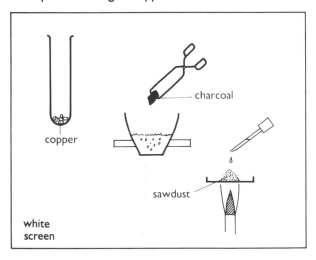

charcoal

copper

sawdust

white screen

What new substance is formed in every test?
Are carbon and copper very reactive elements?
Does copper react with other acids?
Do non-metals such as carbon usually react with acids?

Sodium nitrate and sulphuric acid react. Drops of oily liquid form in the condenser. They collect in the test tube. This is colourless nitric acid. Some brown gas may also form. It dissolves in the nitric acid. It gives the acid a yellow colour.

Copper and carbon are not reactive. Both react vigorously with nitric acid. Sawdust may burst into flame. All three give brown nitrogen dioxide gas. The acid is losing oxygen.

$$4HNO_3 \rightarrow 2H_2O + 4NO_2 + O_2$$

Nitric acid is able to oxidize. It oxidizes charcoal to carbon dioxide. Sawdust burns in oxygen from the acid. Copper may be oxidized to copper(II) oxide. This can then react with the acid to form a salt.

You can see why corks and rubber stoppers are not used. They would react with the nitric acid formed. Another all-glass apparatus is shown. It can be used instead of flask and condenser.

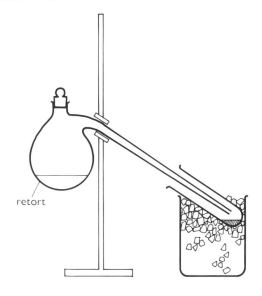

retort

Nitric acid is an oxidant as well as an acid.

Experiment 15.9 Dilute nitric acid and nitrates

Into separate test tubes put
a) magnesium ribbon, pieces of copper, zinc lumps;
b) black copper(II) oxide, zinc oxide;
c) copper(II) carbonate, marble, sodium carbonate.

To each of the eight substances add dilute nitric acid. If there is no change with cold acid, heat gently. Test any gas with lighted splint and with damp litmus paper.

Do any of the metals give off hydrogen?
Which substances give the gas carbon dioxide?
Are any of the products coloured?
Is brown nitrogen dioxide a product of reaction?
Is any unknown gas formed?

Magnesium and cold dilute acid give hydrogen. The mixture becomes very hot. Copper reacts when heated. So does zinc. A colourless gas is formed. It is not hydrogen. It becomes brown at the mouth of the tube.

Copper compounds give blue solutions. They contain copper(II) nitrate. Other nitrates have no colour. There are four ways of manufacturing nitrates.

dilute ⎫ + metal → nitrate + hydrogen or other gases
nitric ⎬ + metal oxide → nitrate + water
acid ⎭ + metal carbonate → nitrate + water + carbon dioxide

The fourth method uses alkali and acid.

Grace: What does a policeman on night duty earn?
George: What does he earn?
Grace: Copper nitrate!

Experiment 15.10 To make sodium nitrate crystals

Use the method of Experiment 6.10. Use dilute nitric acid in the burette. Add it to sodium hydroxide solution. Keep the crystals for later use.

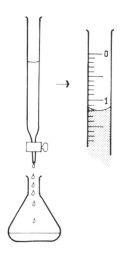

We need a test to recognize a nitrate.

Experiment 15.11 The brown ring test

Put solutions of several nitrates in separate test tubes. Include very dilute nitric acid. Add to each iron(II) sulphate solution. Tilt each tube. Pour in concentrated sulphuric acid. Let it run to the bottom.

What result is the same in every test tube?

The sulphuric acid runs to the bottom. It forms a separate layer. A brown ring forms above the acid. It proves a nitrate is present.

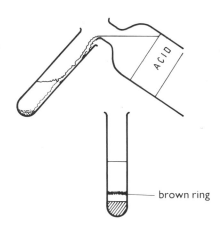

brown ring

Heating nitrates

Experiment 15.12 To heat sodium nitrate

a) Use the crystals from Experiment 15.10. Put them in a small test tube. Hold it in a holder. Heat gently and then strongly. Test any gas with a lighted and then a glowing splint.

Are the crystals easy to melt?
What result suggests that a gas is given off?
Do the splints show what this gas is?
Did you see other changes during the heating?
Is sodium nitrate easy or hard to decompose?

b) **Heat and test other nitrates in the same way. Some contain water of crystallization. Genly heat to drive this off as steam. Then heat the solid nitrate. Compare results for potassium, magnesium and copper nitrates.**

Sodium nitrate is not easy to melt. It needs strong heating. Very strong heating is needed to decompose it. Bubbles appear. The gas is oxygen. The liquid becomes pale yellow. It cools to an off-white solid, sodium nitrite, $NaNO_2$. It is a salt of nitrous acid.

Does potassium nitrate decompose in the same way?
Is it harder or easier to decompose?
Is potassium above sodium in the Activity Series?

Copper nitrate crystals give steam.
What gas is seen when the nitrate decomposes?
Does the gas contain oxygen?
A black powder is left after heating. What is it?
Is the nitrate easy to decompose?
Which needs stronger heating, magnesium or copper nitrate?

Potassium nitrate and sodium nitrate decompose in the same way. Both need very strong heating. Potassium nitrate is less easy. Copper is low in the Activity Series. Its nitrate is easy to decompose. Black copper oxide forms. Brown nitrogen dioxide is seen. It is mixed with oxygen. Magnesium nitrate gives the same gases.

*c) Heat a little lead nitrate in the apparatus shown. Pass the gases through the cooled tube. Collect any gas which comes through. Test it with a glowing splint.

What substance remains in the tube?

Nitrogen dioxide condenses to a liquid in the cold tube. Oxygen bubbles through the water. The other product is solid lead(II) oxide.

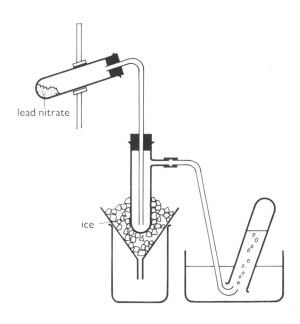

lead nitrate

ice

Heat decomposes nitrates. Nitrates of active metals form the nitrite and oxygen.

$$2NaNO_3 \rightarrow 2NaNO_2 + O_2$$

Other nitrates split up more easily. Nitrogen and oxygen are formed.

$$2Pb(NO_3)_2 \rightarrow 2PbO + 4NO_2 + O_2$$

The solid metal oxide is the other product. Mercury and silver oxides decompose too.

$$Hg(NO_3)_2 \rightarrow Hg + 2NO_2 + O_2$$

The final product from these nitrates is the metal.

The uses of nitric acid

About three million tonnes of nitric acid are made in Britain every year. Most of it is used to make nitrates. They are important fertilizers. Nitrates provide plants with nitrogen. Ammonium nitrate is a fertilizer and an explosive. Potassium nitrate is used in making gunpowder. Sodium nitrate is used as a fertilizer.

Nitric acid reacts with carbon compounds. It adds nitro groups, $-NO_2$. This change needs concentrated sulphuric acid as well. They react with the cellulose in cotton. The result is gun cotton. They react with methylbenzene (toluene). The result is TNT.

Both substances are explosives. An explosion is a very fast reaction. It is over in a fraction of a second. The products are gases. Heat is given out in large quantities. The hot gases exert a huge pressure. They make room for themselves. Things around them are 'blown up'. Most explosions must be started. Detonators and heat can do this.

Summary

Ammonia burns in oxygen. With a catalyst it also reacts with oxygen in air. Nitrogen monoxide forms. On cooling this reacts with oxygen. Brown nitrogen dioxide is formed. This gas dissolves in water. The solution contains two acids. One is nitric acid HNO_3. The other is nitrous acid, HNO_2. Oxygen from air oxidizes HNO_2 to HNO_3.

These changes are used to manufacture nitric acid. Ammonia is mixed with large amounts of air. The mixture passes through platinum gauze at 900 °C. The nitrogen dioxide is mixed with more air. It is then dissolved in water. Nitric acid is the product.

The salts of nitric acid are nitrates. They are made from the acid by four normal methods. They can be recognized by the 'brown ring' test. All nitrates decompose on heating. They do so in three main ways.

Nitrates of active metals give the nitrite and oxygen. Other nitrates give nitrogen dioxide and oxygen. Most of them leave a residue of the oxide. Some decompose to leave the metal itself.

Nitric acid oxidizes. It gives up oxygen. Brown nitrogen dioxide is one product. This shows that oxidation is taking place. The acid can react with the less active metals. Copper gives copper(II) nitrate.

Nitric acid is very important. It is used to make nitrate fertilizers. These provide nitrogen. It is also used to make explosives such as TNT.

Questions on nitrogen and carbon chemistry

1. Nitrogen can be made from air. The apparatus for doing it is shown.

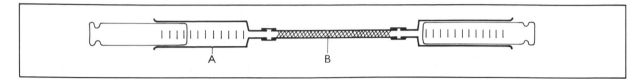

A contains 100 cm³ of air. The substance in B is heated. The air is pushed through it several times. At the end the volume of gas left in A is 79 cm³. It is tested.
   a) Name vessel A.
   b) What is the substance in B?
   c) What colour is it at the start?
   d) What colour does it become when heated?
   e) What gas does it take out of the air?
   f) What is the percentage of this gas in air?
   g) Name the chief gas left in A at the end.
   h) Name another gas present in small amounts.
   j) What do these gases do to a lighted splint?

2. A powdered food is heated with soda lime. It blackens. Smoky vapour comes off. This smells like smelling salts. It turns red litmus blue. Mixed with hydrogen chloride it gives white smoke. This smoke settles as white crystals.
   a) What is the black substance likely to be?
   b) Is the food a carbohydrate, a fat or a protein?
   c) Name one foodstuff which would give these results.
   d) What do the tests with litmus show?
   e) Only one common gas gives these test results. What is it?
   f) Give the formula of this gas.
   g) What ions in solution turn litmus blue?
   h) Name the substance in the white crystals. Write an equation for the reaction which gives them.

3. Ammonia is made on a large scale by this reaction
$$N_2(g) \quad + \quad 3H_2(g) \rightleftharpoons 2NH_3(g) \quad \Delta H \text{ is } -$$
   substance A   substance B
   a) Name substances A and B.
   b) What does the sign $\rightleftharpoons$ mean?
   c) ΔH is $-$. Is the reaction exothermic?
   d) How does heat affect the speed of a reaction?
   e) What is a catalyst?
   f) Which substance is a catalyst for this reaction?
   g) What does (g) mean in the equation?
   h) What volume of B reacts with 10 cm³ of A?

j) Why are A and B compressed before reaction?
k) As the products cool what happens to the ammonia?

4. A platinum coil is made red hot. It is put into the flask as shown. It continues to glow red hot. Brown gas is formed.
   a) What two substances are reacting?
   b) For what reason is platinum used?
   c) Why does it stay red hot?
   d) What is the name of the brown gas?
   e) What is its formula?
 · f) What effect will the brown gas have on litmus?
   g) What substance forms in water when the brown gas dissolves?
   h) What substance is made on the large scale by these reactions?

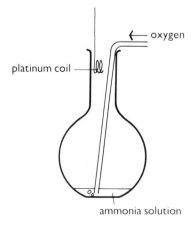

5. Give one important use of each of the following substances.
   a) ammonia,
   b) ammonium chloride,
   c) ammonium sulphate,
   d) nitric acid,
   e) ammonium nitrate,
   f) potassium nitrate.

6. Nitrates are made from nitric acid. Name the substance you would add to the acid to make
   a) sodium nitrate,   c) copper(II) nitrate,
   b) ammonium nitrate,   d) magnesium nitrate.

7. Nitrates decompose on heating in three ways:
   a) to form a nitrite and oxygen
   b) to form a metal oxide, nitrogen dioxide and oxygen.
   c) to form a metal, nitrogen dioxide and oxygen.
   In which way, a), b), or c) do these nitrates decompose:
   i) lead nitrate,
   ii) copper nitrate,
   iii) potassium nitrate?
   Explain how you would separate the two gases formed in b).

8. Choose the best answer in each case:
   a) The percentage of nitrogen in air is
      A 0.03%   C 21%   E 98%
      B 1%   D 78%
   b) The formula of nitric acid is
      A NO_2 C HNO_2 E NH_3
      B NO D HNO_3
   c) Soil in a garden has just been limed with calcium hydroxide. It would be wrong to add one of the following fertilizers because it would react with the lime. Which one?
      A sodium nitrate
      B potassium nitrate
      C ammonium sulphate
      D sodium chloride
      E chalk
   d) Esters are formed by the reaction between
      A acid and alkali
      B alcohol and oxidizing agent
      C sugar and water with yeast present
      D alcohol and organic acid
      E monomer molecules of the same substance
   e) The formula of ethane is
      A CH_4, B C_2H_4, C C_4H_{10}, D C_2H_6, E C_2H_2

9. Give the name of each compound shown below as a formula:
   a)
   H
   |
   H-C-H
   |
   H
   b)
   H H
   | |
   H-C-C-H
   | |
   H H
   c)
   H H
   | |
   H-C-C-O-H
   | |
   H H
   d)
   H H H
   | | |
   H-C-C-C-H
   | | |
   H H H
   e)
   H   H
    \\   /
     C=C
    /   \\
   H   H
   f)
   H
   |
   H-C-Cl
   |
   H

i) Which formula contains a double bond?
ii) Which of the six is an alcohol?
iii) Three of the six substances belong to the same family or homologous series. Which ones?
iv) Substance e) is shaken with bromine water. What would you see? Use this reaction to explain 'addition reaction' and 'unsaturated compound'.
v) Use the formulae a) and f) to explain what is meant by a substitution reaction.
vi) Which of the six substances can be used to form an ester?

10. Describe how you would make a small sample of
    a) soap, b) detergent in the laboratory.
    Explain why hard water affects soap. Does a detergent affect hard water in the same way?

11. Crude oil is distilled in a refinery. It takes place in a tall tower. Liquids are piped off at different heights up the tower. Explain how the method works. Use the words 'fraction', 'bubble caps', 'boiling point', 'gasoline'.

12. $CH_4 + 2O_2 \rightarrow CO_2 + 2H_2O$ $\Delta H = -888$ kJ per mole
    Using the equation answer the following questions:
    a) Name the substance with formula CH_4.
    b) Is it an alkane or an alkene?
    c) How many molecules of oxygen does one molecule of CH_4 react with?
    d) What are the products of reaction?
    e) Is the reaction exothermic or endothermic?
    f) What does ΔH stand for?
    g) What does kJ stand for?
    h) What is one mole of CH_4? (C = 12, H = 1)
    i) Draw a structural formula for CH_4.
    j) Is it saturated or unsaturated?

13. Explain each of the following terms:
    a) monomer,
    b) polymer,
    c) isomers,
    d) homologous series,
    e) enzyme,
    f) fermentation.
    Give one example of each in your explanation.

14. The formula of propanol is C_3H_7OH. Draw two different structural formula for this formula. Name the two propanols they represent. What is this effect called?

Fertilizers

Human beings need food. It gives us energy. It provides us with growth. Our bodies use it for repair and replacement of body tissue.

We grow plants and breed animals to use as food. The animals eat plants too. So all our food supply depends on plants. We need to grow them well.

How big is a carrot seed? How big is the carrot it grows into? Where does all the extra stuff come from?

What do plants need for growth?

Plants need carbon, hydrogen and oxygen. They get them by taking in carbon dioxide and water. Using energy from sunlight they make sugar. The sugar is built up into starch and cellulose. These become the stem, roots and leaves of the plant. This process is called photosynthesis. Only a water shortage will stop photosynthesis.

Plants must have nitrogen, phosphorus and potassium. From nitrogen plants build protein. Potassium helps them to do this. Phosphorus promotes general growth. Most plants use compounds of the three elements.

Plants must also have compounds of calcium, sulphur and magnesium. Tiny amounts of other elements are needed. They are called 'trace elements'. Among them are iron, copper and manganese.

People and animals rely on plants for food.

How does the plant crop get trace elements?

Plants use compounds of all these elements. They are present in most soils. They dissolve in water. The plants take them in through their roots.

In nature, plants grow, seed and die. They decay and break down into compost. This contains all the substances the plants took in. They pass back into the soil. The next lot of plants can use them.

The crops we grow are eaten. Human beings and animals use them as food. The soil loses some compounds. Manure puts some of them back but not all. The soil becomes poorer. It will not grow such good crops. Decayed vegetable matter (humus) is also needed.

Chemists can manufacture the compounds. They can be added to the soil. They are called fertilizers. They must dissolve in water in the soil. Then they make up for those lost. Fertilizers do not supply humus.

The common nitrogen fertilizers are nitrates and ammonium salts. Others are liquid ammonia and urea. Urea, $CO(NH_2)_2$, is made from ammonia.

Phosphates, salts of phosphoric(V) acid, supply phosphorus. The calcium salt occurs in nature. It is called 'phosphate rock'. It does not dissolve well. Sulphuric acid converts it into 'superphosphate'. This dissolves well, so that plant roots can take it in.

Potassium chloride, nitrate and sulphate supply potassium. Most fertilizers we buy are mixtures. They contain compounds of the three main elements. Some trace elements are there as well.

Moving bags of fertilizer in a warehouse using a robot machine.

The Third World

Over 4000 million people live on the earth. One in eight of them is near to starving. Over half of them are without one kind of food or another. This is called malnutrition. People who suffer from it need our help. Most of them live in poorer countries. These people are often called 'The Third World'.

Richer countries can help in three main ways.
They can increase the world food supply.
They can ensure that it is shared out more fairly.
They can teach poorer people how to improve their crops.

Third World crops can fail because of water shortage.

Their crops fail for three main reasons. They suffer from water shortage or poor soil. The plants die of disease. Crops are eaten by pests. The growing plants may be eaten. Pests may also eat the crops after harvesting.

Sinking wells and building dams will help. Dams collect water in the rainy season. They store it for times of drought. Better types of seed and livestock help. They give bigger food yields. The use of fertilizers improves crops still more.

Chemical substances can prevent crop disease. Others can destroy the pests which eat the crops. Chemical sprays save crops. They can also have harmful effects. They may harm people. They kill useful insects.

Crops are sometimes destroyed by pests. These maize plants have been damaged by locusts.

Food and you

We all eat food. It is a necessity and a pleasure. We use it for four main purposes. First, it gives us energy. We found out about this on page 52. Secondly, we need food for growth. It is also used to repair and replace body tissue. Lastly, some foods control all these changes.

Foods belong to several classes.
Carbohydrates contain carbon, hydrogen and oxygen. They burn to give energy. One example is sugar.
Fats contain the same elements. They come mostly from animals. Meat and butter are examples. Fats provide energy too. Our bodies store energy as fat.
Proteins are nitrogen compounds. Some contain sulphur and phosphorus. They are present in body tissue. We need compounds of certain elements. They are called **mineral salts.** Good diet provides them. The last class is **vitamins.** They are essential to life. They are labelled A, B, C, D and E.

We grow plants and animals to use as food. Animals eat plants too. So all our lives depend on plants. Without them we starve and die. Tens of millions of people do this every year.

Green plants are factories. A factory manufactures. It starts with 'raw materials'. It makes 'products'. Plants use water and carbon dioxide. They manufacture starch and sugar. This is photosynthesis (page 55). Starch is converted into cellulose. All three are carbohydrates. They make up the body of a plant.

Most food plants grow in soil. Their roots push down into it. They take water from it. The roots also anchor the plant. The soil must allow the roots to do both.

Soil consists of tiny solid particles. They come from the breakdown of rocks. But natural plants are not harvested. They die and decay. The plant remains pass into the soil. This provides humus. Humus is the peaty part of soil. It holds water and air.

Soil is also the home of animal life. It ranges from large worms to tiny mites. A handful of soil holds millions of other organisms. They include many kinds of bacteria. They 'work' the soil. They keep it

Foods rich in carbohydrate.

Foods rich in fats.

Foods rich in protein.

porous. They also produce compounds which plants can use. Nature makes good soil in this way. It does so slowly. A 12-inch depth may take a 1000 years to make.

The compounds include mineral salts. They dissolve in water in the soil. The roots take in this water. The plant gets essential salts. Plant decay puts them back into the soil.

We grow plants. At harvest we take them up. The soil loses. Less humus is put back by decay. We have to replace it. Manure and compost do this. They also put back mineral salts. So do fertilizers (page 246). What is needed for good crops?

Good soil is needed to grow them in. Good soil has water and air. It has humus and mineral salts. They must be in the right proportions. Then plants will grow. The soil is fertile. How much of the earth has fertile soil? Only about one tenth of it.

Some is too dry. Rain falls rarely, if ever. Over a quarter of the earth is desert or suffers from drought.
Some is too wet. Sodden soil contains no air. Some is short of mineral salts. Some areas are too cold. Some are too hot. Plants do not begin to grow in them.

Soils can deteriorate. Over-cropping can do this. It takes out of the soil. It does not put back. Without humus soils dry out. They become dusty. Winds blow soil away. Flooding can wash it away. Towns and house building can use it up. Soil is life!

The world has a severe food problem. Some countries have more than they need. Others have too little. What can be done? We must:

Find methods of preserving good soil
Teach and help people to use it better
Find ways of distributing surplus food
Develop other sources of food. Vegetable oils can be turned into edible fats. Proteins can be made by chemical reaction. Petroleum may be suitable raw material.

All our lives depend on plants.

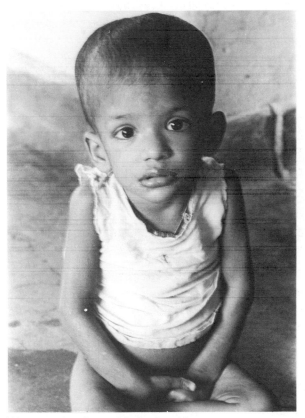
We must solve the world's food problem.

247

How fast is a reaction?

In a chemical change substances react. They are the reactants. Their atoms and molecules form new substances. These are called products.

reactants → products

Many reactions are very slow. A piece of iron may take years to rust. Some reactions are very fast. An acid and alkali react the instant they meet. We prefer to slow down reactions like iron rusting. It is useful to be able to speed up many others.

The speed of a reaction is called its **rate**. We can measure it. We often measure the rates of other changes. How? We find out how some quantity alters. What is your rate of spending? How much money did you start with? How much has gone? In how many days?

at 8 a.m. on Monday	80p
at 8 a.m. on Friday	20p
money spent	60p
time taken	4 days
rate of spending	$\frac{60}{4}$
rate =	15p per day

A bucket has a hole in it. How fast does it leak?

Method 1. Mark the bucket in litres. Fill it. Start a stop watch. Every minute note the volume left. Work out the rate of leaking. It will be only an average. It is measured in litres per minute.

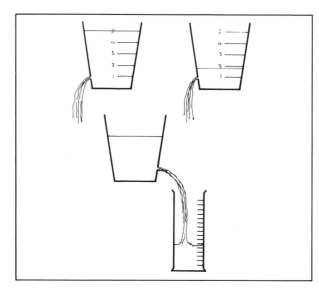

How fast is a reaction?
Measure one reactant in moles. It will steadily be used up. At intervals measure how much is left. **Rate of reaction** will be in **moles** used **per second.**

Method 2 is better. Fill the bucket. Note the time. Catch the water produced. At intervals note the amount collected. Water is a product of the leak.

Reactions form products. Collect and measure one of them. **Rate** will be in **moles** formed **per second.**

Some factors alter the rate of reaction. Method 2 can show this. Collect the product in every case. Compare the amounts formed in the same time.

Experiment 16.1 The effect of size when a solid reacts

Measure exactly 20 cm³ of 2 M hydrochloric acid. Put it in a conical flask. Fit a stopper and delivery tube. Select large marble lumps of about the same size. Take 20 g of them. Pour them into the flask.

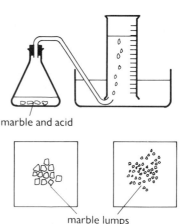

marble and acid

Stopper it at once. At the same moment start a stop watch. Collect the gas product in a measuring cylinder. Note the volume of gas every 30 seconds.

marble lumps

Take more of the marble lumps. Break them into smaller pieces. Take 20 g of these. With them do the same experiment again. Plot both sets of results as a graph.

One product of the reaction is carbon dioxide. The graph shows my results. Times are on the horizontal axis. From the graph we can get the volumes of gas formed after 30 seconds. Use *your* graph if you can.

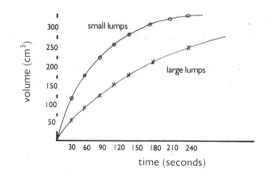

What volume did the large lumps give after 30 seconds?
What volume did the small lumps give after 30 seconds?
Which lumps give the faster reaction?
Can you explain why?

Acid and marble react only where they meet. This is at the marble surface. The smaller lumps give gas more quickly. With them rate of reaction is greater.

The drawing shows why. Cutting a large lump gives a greater surface. More marble and acid will meet.

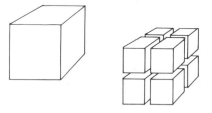

Catalysts

A catalyst is a substance added to a reaction to alter its rate. It is still there at the end.

Experiment 16.2 The action of catalysts

Colour 10-volume hydrogen peroxide with Universal Indicator. Add drops of alkali until the colour is green. The hydrogen peroxide is then neutral.

a) Half fill four test tubes with the neutral solution. Stand them in water in a beaker. To tube 1 add dilute acid. To tube 2 add the same volume of alkali solution. Add the same volume of water to 3.

Heat the beaker. Watch the tubes. Compare how fast gas forms in them. Add to 4 a pinch of manganese(IV) oxide. Test the gas given off with a glowing splint.

What gas does the reaction give?
How fast is this product formed in tube 3?
Is the rate faster in acid solution in tube 1?
Does alkali increase the rate in tube 2?
Does manganese(IV) oxide act as a catalyst?

b) Half fill six tubes with hydrogen peroxide. Add different amounts of manganese(IV) oxide to three. Try other metal oxide powders in the other three.

Does the amount of catalyst used make a difference?
Are any other oxides catalysts for this reaction?

Test tube 3 is a **control** experiment. The hydrogen peroxide is neutral. It has no catalyst in it. Its reaction is normal. It decomposes giving oxygen. The stream of bubbles shows the rate of reaction. We can see what factors alter the rate.

Manganese(IV) oxide and alkali increase it. They are catalysts for this reaction. With acid no gas forms. Acid decreases the rate. Heating always increases the rate.

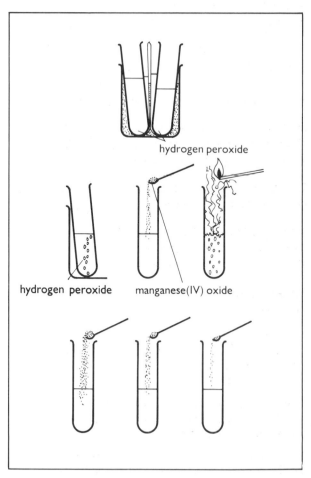

Concentration and temperature

Marble and acid form a gas. We *collect* it. The results show best on a graph. Hydrogen peroxide also gives a gas. We can *see* how fast it is formed. So we are able to compare rates of reaction.

Then we can show which factors alter the rate. In hydrogen peroxide acid does. It decreases the rate. It is called a **negative catalyst**.

Metal oxides are catalysts for this reaction. Manganese(IV) oxide is much the best. The more catalyst used the greater its effect.

Experiment 16.2 also shows the effect of heat. Gas comes off faster. The rate is increased. We watered down the solution in test tube 3. Will this alter the rate of reaction?

Experiment 16.3 The effect of concentration

Take 100 cm³ of distilled water. Add a pinch of sodium hydrogencarbonate. Put in 12 g of sodium thiosulphate(VI). Stir to dissolve the crystals.

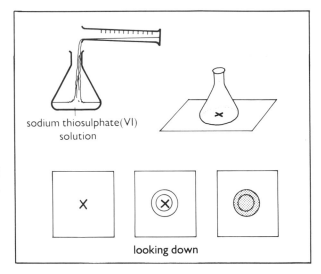

sodium thiosulphate(VI) solution

looking down

Draw a pencil cross on paper. Measure 20 cm³ of the sodium thiosulphate(VI) solution. Put it in a small conical flask. Stand it over the cross. In a measuring cylinder put 20 cm³ of 0.5 M nitric acid. Pour it into the flask.

At the same moment start a stop watch. Swirl the solution in the flask. Look down through it at the pencil cross. When it disappears, stop the watch.

Take a second flask. In it put 10 cm³ of thiosulphate(VI) solution. Add 10 cm³ of distilled water. Pour in 20 cm³ of the 0.5 M nitric acid. Start the stop watch at the same time. Swirl the solution. Find the time taken to blot out the pencil cross.

What product appears in the mixture?
Why does the pencil cross disappear?
In which flask is rate of reaction greater?

time for first flask   23 secs.
time for 2nd flask   47 secs.

Acid and thiosulphate(VI) react. One product is sulphur. It blots out the cross. It does this in both flasks. It needs the same amount of sulphur each time. The mixture in the second flask takes twice as long. The rate has been halved. The solution in this flask was weaker. Its concentration was halved. The greater the concentration the greater the rate.

Experiment 16.4 The effect of temperature

Take two flasks. Put 5 cm³ of thiosulphate(VI) solution in each. Add 15 cm³ of distilled water. Use the first one to do Experiment 11.3. Take the temperature of the solution. Heat the other solution to about 40°C. Do the experiment. Take the temperature.

sodium thiosulphate(VI) solution

What time should you expect for the first flask?

Here are some results.

Temperature	20°C	30°C	40°C
Time taken in seconds	36	18	?

Is reaction faster at higher temperatures? What time would you expect at 40°C?

The first temperature was 20°C. This was the temperature of the reactants. They produce sulphur. They need only thirty-six seconds to produce enough sulphur to blot out the cross.

Then the reactants were at 30°C. They produce the same amount of sulphur. They take only eighteen seconds. They do it twice as fast. The rate of reaction is twice as great.

We can alter the rate of reaction in four ways.

1 We can alter the temperature. A rise in temperature increases the rate. A rise of 10°C often doubles it.

2 We can find a catalyst. Most of them increase the rate. Some, called negative catalysts, decrease it.

3 Small lumps of solid react faster than larger ones.

4 Liquids and gases react. Rate depends on concentration. The greater the concentration the greater the rate.

George drives the mower, grass falls in it.
To fill the box takes just a minute.
By crazy driving Willie reckons
To fill this box in thirty seconds.
This is a simple way of showing,
Who has the greater rate of mowing.

Summary

Reactions vary from very slow to very fast. The speed of a reaction is called its rate. We can measure it in moles per second. We can also compare rates of reaction. The best method is to use one of the products formed. By comparing rates we show that:
a) Small lumps of solid react faster than large ones.
b) A catalyst can be used to increase the rate. Negative catalysts decrease it.
c) A rise in temperature increases reaction rate. A 10°C rise often roughly doubles the rate.
d) Concentration affects liquid and gas reactions. The higher the concentration the greater the rate.

Question

20 g of metal lumps react with 20 cm³ of 2 M acid. Hydrogen is formed. The total volume is measured every minute. The results are shown below.

Draw a graph of these results. Plot time on the horizontal axis. On the graph sketch the results you would expect if
a) the temperature was about 10°C higher;
b) 20 cm³ of M acid were used instead;
c) 20 g of smaller lumps were used in the same acid at the same temperature.
d) The last two volumes shown in the results are the same. Why is this?

Time from start of reaction (mins)	1	2	3	4	6	8	10	12	16	19
Volume of gas formed (cm³)	48	87	118	142	172	203	220	231	240	240

Metals

A metal is easy to recognize. It has metallic lustre. If not, it can be polished up. A metal conducts electric current and heat. Most metals have a high melting point. Most have high densities.

These are physical properties. On page 71 we tested some reactions of metals. We used the results to put the metals in order. The list is called the Activity Series. The most active metal is at the top. The least active comes last. The Series included hydrogen.

Try some of these tests again. Remember, * means 'for teachers only'.

Experiment 17.1 Metals and air

a) Use a bunsen with its hole half open. Hold a strip of metal in the flame. Use copper, zinc, aluminium, magnesium, tin, iron and platinum.

*b) Hold a tiny piece of sodium in tongs. Put it in the flame. Use calcium too. Let the product cool.

Drop each cold product from a) and b) into a test tube. Shake it with distilled water. Add Universal indicator. If the product does not dissolve, add dilute hydrochloric acid. Heat if necessary.

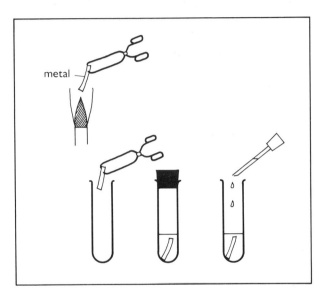

> Which metals burn?
> What are the products?
> Do these products form alkalis in water?
> Which metals form new products without burning?
> Do these products give alkalis in water?
> Do the products dissolve in dilute acid?

Some metals burn in air. They form oxides. These oxides dissolve in water. They react to form alkalis. The alkalis are metal hydroxides. Less active metals do not burn. Many become coated with an oxide. These oxides do not react with water. They do react with acids.

Put the metals in order of how well they react with air.

Experiment 17.2 Metals and acid

Use the same metals as in Experiment 17.1a). Drop each into dilute hydrchloric acid. Warm if needed. Test any gas. Put the metals in order again.

> Is the order the same as in Experiment 17.11?
> Which reactions give heat?
> Does this help to place the metals in order?

Magnesium reacts at once. The mixture becomes very hot. The metal disappears. Hydrogen is given off. Less active metals give less heat. Some need heat to start the reaction. Copper and platinum do not react.

Experiment 17.3 Metals with water or steam

Put a 2 cm depth of water in a test tube. Push into it some rocksill fibre. Set it up as shown. Put a small pile of metal powder in the tube. Heat it. Collect and test the gas. Use copper, iron and zinc.

*Use small piles of magnesium and aluminium.

*Put safety screens round a trough of water. Drop in a small freshly cut piece of sodium. Drop a piece of calcium into a beaker of water. Collect the gas. Test both solutions for alkali.

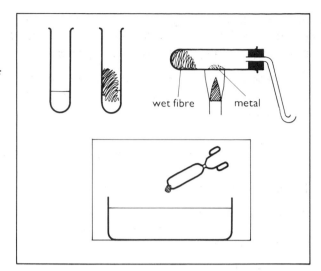

wet fibre   metal

Which metals react with cold water?
Which metals burn in steam?
Which do not burn but become red hot?
Write word equations for all reactions.

Sodium and calcium react vigorously with water. Magnesium burns violently in steam. It also reacts slowly with water. Aluminium burns in steam. Zinc and iron glow red hot. The gas is hydrogen. Typical equations are

calcium + water → calcium hydroxide + hydrogen
zinc + steam → zinc oxide + hydrogen
Both reactions are exothermic. ΔH for each is minus.

Metals which react are turning into their ions. A full equation shows how. It gives in detail each ion formed.

Magnesium burns in air. White magnesium oxide forms.

$$Mg(2.8.2.) + O(2.6.) \rightarrow Mg^{2+}(2.8.) + O^{2-}(2.8.)$$

We may not need all this detail. If not, we write

$$2Mg + O_2 \rightarrow 2MgO$$

We can treat other reactions in the same way.

calcium + water → calcium hydroxide + hydrogen

$$Ca\,(2.8.8.2.) + 2H_2O \rightarrow Ca^{2+}\,(2.8.8.) + 2OH^- + H_2$$

With less detail:

$$Ca + 2H_2O \rightarrow Ca\,(OH)_2 + H_2$$

sodium
calcium
magnesium
aluminium
zinc
iron
tin
(lead)
hydrogen
copper
platinum

Metals react with air, acid, water and steam. Each puts the metals in order. The order is roughly the same in each case.
Copper reacts only with air. It cannot take oxygen from steam. On page 64 we heated copper(II) oxide in hydrogen. Hydrogen is more active than copper.

$$\text{copper oxide} + \text{hydrogen} \rightarrow \text{copper} + \text{steam}$$
$$CuO + H_2 \rightarrow Cu + H_2O$$

My name is William Henry Lamb.
It says exactly who I am.
Unless I need to be official
I may use only one initial.
With those who know me well its silly
To use more detailed names than Willie.

copper(II) oxide, $Cu^{2+}.O^{2-}$

CuO

copper oxide

253

Metal versus metal again

Elements fall into two classes, or groups. An element is either a metal or a non-metal. Its properties decide which. Putting things into classes is called classification.

All metals have some similar properties. But some are more active than others. We place them in order of how active they are. This is a second kind of classification. The list is the Activity Series. We can use it to predict results. It is also called the Reactivity Series.

People who do football pools predict. Before the games they say what the results may be. Magnesium is more active than copper. Suppose we mix magnesium with copper oxide. Will they react?

Experiment 17.4 Metal versus metal

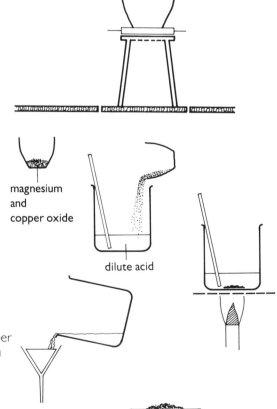

magnesium
and
copper oxide

dilute acid

*a) Dry magnesium powder and copper(II) oxide in an oven. Mix small amounts of them in a crucible. Stand the crucible on a pipe clay triangle on a tripod. Stand the tripod on tiles. Put safety screens around it. Light a bunsen burner under the crucible.

Did you predict that the two substances would react?

The reaction is almost explosive. Nothing is left in the crucible. We can only guess what is formed.

*b) Put some of the magnesium and copper oxide in another crucible. Mix in magnesium oxide. Heat the crucible. Let it cool. Scrape the contents into a beaker. Add dilute acid. Heat to dissolve the white oxide. Filter.

What substance is left in the beaker?
Write a word equation for the crucible reaction.

Adding magnesium oxide makes the reaction less vigorous. The products are not blown out of the crucible. Acid dissolves magnesium oxide. We can see the other product. We can prove that it is copper.

magnesium + copper oxide → magnesium oxide + copper
Mg + CuO → MgO + Cu

We could predict this result. Magnesium is much higher in the Series than copper. It is much more active. It can take oxygen from copper oxide. It will do so vigorously.

254

c) Mix copper(II) oxide with iron powder. Put a small pile of the mixture in a porcelain boat. Wear eyeshields. Heat the mixture gently and then strongly. Heat iron with other oxides in the same way. Use zinc oxide, a lead oxide and aluminium oxide. Try to predict each result before you begin.

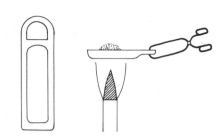

Which mixtures glow when they are heated?
Is the glow a sign of a chemical change?
Which mixtures show no change?
Did you predict each result correctly?

Iron is above copper in the Series. It is more active. It will take oxygen from copper oxide. Taking away oxygen is reduction. Iron reduces lead oxide. It does not affect zinc and aluminium oxides.

iron + copper(II) oxide → iron oxide + copper

Iron has 'displaced' copper. Can it do this with other copper compounds. What other copper compound could we try?

Experiment 17.5 Displacement reactions

Clean pieces of iron, copper and zinc with emery paper. Dip iron into copper(II) sulphate solution. Dip zinc into lead(II) nitrate solution. Put the copper into silver nitrate solution. After a few minutes take them out. Observe the results.

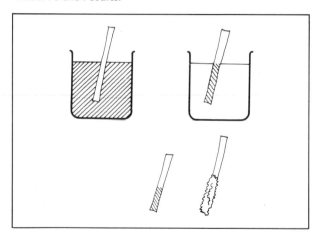

What metal forms on the copper foil?
Where should silver appear in the Activity Series?
What forms on the surface of the zinc?
Does iron always displace copper?

Each metal becomes coated. Copper appears on the iron. Crystals of shiny lead appear on the zinc. A white metal is formed on the copper. This is silver. Silver comes below copper in the Series.

Oxidation and reduction

Using the simple formulae for the reaction

$$Mg + CuO \rightarrow MgO + Cu$$

Magnesium gains oxygen. $Mg \rightarrow MgO$ is oxidation. Copper oxide loses oxygen. $CuO \rightarrow Cu$ is reduction.

We can write the equation to show the ions.

$$Mg + Cu^{2+}.O^{2-} \rightarrow Mg^{2+}.O^{2-} + Cu$$

A magnesium atom gains oxygen. It turns into an ion. Mg(2,8,2.) loses 2 electrons. It becomes Mg^{2+} (2.8.) Gaining oxygen is the same as losing electrons.

$Cu^{2+}O^{-} \rightarrow Cu$. A copper ion turns into an atom. To do this it gains 2 electrons from an Mg atom.
Loss of oxygen = Gain of electrons

Oxidation is gaining oxygen or losing electrons. Reduction is losing oxygen or gaining electrons.

If one substance gains oxygen another must lose it. One substance loses electrons. Another gains them. Oxidation and reduction occur together. They happen in the same reaction. It is called a **redox reaction**. Redox comes from **red**uction-**ox**idation.

255

The extraction of metals

Metal compounds are found in the earth. They are called ores. Getting a metal from its ore is called extraction. There are many methods of doing this.

Mining bauxite.

Most ores are oxides, carbonates or sulphides. They are not pure. They contain rock and other substances. These make extraction much harder. The first step is to remove them. As much impurity as possible is taken out. This is called refining the ore. It makes it much purer.

Ores are metal compounds. So they contain ions of the metal. Extraction turns ions into atoms. The ions gain electrons. This change is reduction. Some ores are easy to reduce. Others are difficult.

The Activity Series tells us which. The top metals react vigorously. Changing their atoms to ions is easy. So the reverse change will be difficult. Ores of the top metals are hard to reduce to the metal.

Metals lower down react less easily. Turning their atoms into ions is harder. Then their ores will be easier to reduce. Extraction of these metals is less difficult.

The least active metals are copper, silver and gold. They are even easier to extract. They also occur as the metals themselves. They occur 'native'. So they were known and used very early in history.

Metals are extracted in huge amounts. The simplest method is to reduce the oxide. The best reducing agent is carbon. Coke is the form usually used. It is easily made from coal. It is fairly cheap. As it reduces the ore it forms carbon oxides. These need not be removed. They escape as gases.

Many ores are metal oxides. Carbonate and sulphide ores can be converted into oxides. They are strongly heated in air, or 'roasted'. For example, calamine and zinc blende are zinc ores. Calamine is zinc carbonate. Zinc blende is the sulphide. Both are roasted to give zinc oxide.

$$ZnCO_3 \rightarrow ZnO + CO_2$$
$$2ZnS + 3O_2 \rightarrow 2ZnO + 2SO_2$$
$$\text{(sulphur dioxide)}$$

The oxide is mixed with coke and heated.

The temperature needed varies with the metal. For metals high in the Activity Series it is very high. It can be over 2000°C.

Manufacturing coke.

High temperatures are very costly to produce
need special furnaces to stand the heat
make it hard to separate the products

For active metals we need an easier method. Electrolysis is used. A suitable compound is chosen. It is often the chloride. It is melted. A current is passed through it.

calcium chloride
Ca^{2+} $2Cl^-$

Metal ions move to the cathode. It is the negatively charged electrode. It provides the ion with electrons. The metal is formed. This cathode reaction is reduction.

at the cathode
$Ca^{2+} + 2e \rightarrow Ca$ (metal)

Chlorine ions reach the anode. They give up electrons. Chlorine atoms form. They combine in pairs. Chlorine gas is given off. The anode reaction is oxidation.

Gain of electrons
at the anode
$Cl^- - e \rightarrow Cl$
$2Cl \rightarrow Cl_2$ (gas)

Why is a melted compound used? Why not a solution? The cathode reaction is the same. Solution is cheaper. Remember we are extracting 'top' metals. They are very active. Many react with water. Calcium does. It would react with water from the solution.

The table shows the methods of extraction. The metals are in Series order. You can see how method links up with position.

Metal	Ore used	Reduction of oxide by carbon	Actual method used
Potassium Sodium Calcium Magnesium	chloride chloride carbonate carbonate	Temperature needed for this reduction is much too high. Magnesium has been made this way.	Electrolysis of the molten chloride. Calcium and magnesium chlorides are made from their carbonates.
Aluminium	oxide	Temperature too high. Aluminium reacts with carbon.	Electrolysis of the oxide dissolved in a molten aluminium salt.
Zinc Iron Tin Lead Copper	sulphide oxide oxide sulphide sulphide	The temperature for reduction is below 1000°C. For copper oxide it is below 100°C	Sulphide ores are roasted to give the oxide. This is mixed with coke. The mixture is heated.
Silver Mercury Gold	sulphide sulphide native	Reduction by coke is not needed.	Extraction with cyanide. Heating the ore→metal. Extraction with cyanide.

The Periodic Table

On page 174 we compared chlorine, bromine and iodine. Their reactions are very much alike. They belong to a chemical family called the halogens. Look at their relative atomic masses.

Cl	Br	I
35	80	127
	45	47

Lithium, sodium and potassium are very much alike. They belong to a family too. They are alkali metals. Their relative atomic masses differ by equal amounts.

Li	Na	K
7	23	39
	16	16

Dobereiner was the first to notice this. He said that properties and relative atomic mass were connected. In 1865, about sixty elements were known. Newlands put them in order of atomic mass. Similar elements came at regular intervals. Newlands said they were like notes in a scale. The first and second doh are not the same note. But they are very much alike.

doh re me fah sol lah te *doh*

In 1869 a Russian took the idea further. His name was Mendeleef. He put the elements in order of relative atomic mass. The smallest one came first.

H Li Be B C N O F Na Mg Al Si P S Cl K Ca . . . .

He built the list into a table.
It is called the Periodic Table.

```
H
Li  Be  B   C   N   O   F
Na  Mg  Al  Si  P   S   Cl
K   Ca  ..  ..
```

On page 158 there is a modern Periodic Table. It has in it 104 known elements. They include the noble gases. These were unknown in 1869.

Each element is shown in the Table by its symbol. Those in black are of the more common elements. The atomic number is shown above the symbol. The name of the element is in small print. Below this is the relative atomic mass.

20	atomic number
Ca	symbol
Calcium	name
40	relative atomic mass

The vertical lines of elements are called groups. They are labelled with Roman numbers. Each group contains a chemical family.

The horizontal rows are called periods. They are numbered from 1 to 8. Metals are on the left-hand side of a period. Non-metals are found on the right.

GROUPS

	I	II	III	IV	V	VI	VII	O
Period 1	H							He
Period 2	Li	Be	B	C	N	O	F	Ne
Period 3	Na	Mg	Al	Si	P	S	Cl	Ar
Period 4	K	Ca	..	..				

Mendeleef had many problems. We shall deal with two of them. We shall use the modern Table.

In 1869 many elements had yet to be discovered.
Suppose silicon, Si, had been one of them.
Without it the list becomes

H He Li Be B C N O F Ne Na Mg Al P S Cl Ar K

The Periodic Table then becomes:

I	II	III	IV	V	VI	VII	O
H							He
Li	Be	B	C	N	O	F	Ne
Na	Mg	Al	*P*	*S*	*Cl*	*Ar*	K
							noble gases

This looks wrong. It puts phosphorus under carbon.
Sulphur appears in Group V under nitrogen.
Argon, a noble gas appears in the halogen family.
The metal potassium is in Group O with the noble gases.

Mendeleef avoided this by leaving a gap. He said that an element would be discovered to fill the gap. He predicted its properties. He also described its compounds in detail.

II	III	IV	V	VI
Be	B	C	N	O
Mg	Al	..	P	S

Each element for which he left a gap was discovered. Its properties were found to match Mendeleef's predictions. Its compounds were as he had said they would be.

The second problem came at element 21. Element 20 is calcium. It fits well in Group II with beryllium and magnesium.

	I	II	III	IV	V		
Period 4	K	Ca	21	22	23	..	..

Element 21 is scandium. It fits fairly well in Group III. The next nine elements fit less and less well. The metal manganese, for example, comes in the halogen group VII.

	I	II	III	IV	V	VI	VII
Period 3	Na	Mg	Al	Si	P	S	Cl
Period 4	K	Ca	Sc	Ti	V	Cr	Mn

It is simpler to see this in the modern Table. Take out the elements from scandium to zinc. This brings us to element 31, gallium, Ga. It fits well in Group III. Those which follow it fit well too. The period ends with krypton, Kr. This noble gas is in its right group, Group O.

I	II	III	IV	V	VI	VII	O
K	Ca	..	..	..	..	..	..
..	..	Ga	Ge	As	Se	Br	Kr

Element 20 is in the right group. It has Group II properties. Elements 31 to 36 are also correctly placed. Each has the properties of its group. The elements in between do not fit into groups. All ten are very much alike. They are all metals. They form two or more different ions. They have coloured compounds. Many act as catalysts. So do their compounds. They are as alike as members of a group. Zinc and scandium are less alike.

II						VIII					III
Ca 20	Sc	Ti	V	Cr	Mn	Fe	Co	Ni	Cu	Zn	Ga 31
					Transition metals						

They are called the transition metals.
Periods 5 and 6 have transition elements. Among them are platinum, silver, gold and mercury.

Modern ideas and the Periodic Table

From 1896 onwards, experiments 'opened up' the atom. They showed that every atom has a nucleus. The nucleus contains protons. The number of protons is called the atomic number of the element. It turns out to be the position of the element in the Periodic Table. It is also the number of electrons in the atom.

Electrons are in shells. The outside shell is important.
When an atom reacts the outside shell loses or gains electrons.
the ions formed attract each other.
or the outside shell shares electrons with other atoms.
covalent bonds hold the atoms together.

How does this fit in with the Periodic Table?

Periods differ in length. Period 1 contains hydrogen and helium, a noble gas. Helium has two electrons. They fill the first shell.

Period 1	H		He
Electrons	1		2

Period 2 has eight elements. It ends, like Period 1 with a noble gas, neon. The outside shell of each atom is the second shell. It is full when it has eight electrons.

Period 2	Li	Be.............................	F	Ne
Electron structure	2.1	2.2	2.7	2.8

Sodium is the first element in Period 3. Its electron structure is 2.8.1. The last element in Period 1 is argon. It is a noble gas, structure 2.8.8.

Group I		II	
Hydrogen	1		
Lithium	2.1	Beryllium	2.2
Sodium	2.8.1	Magnesium	2.8.2
Potassium	2.8.8.1	Calcium	2.8.8.2

Period 4 begins with potassium, 2.8.8.1. Its outside shell has one electron. So have the atoms of sodium and lithium. This is why all three react in the same way.

Li	2.1	$\rightarrow Li^+$	+ 1 electron
Na	2.8.1	$\rightarrow Na^+$	+ 1 electron
K	2.8.8.1	$\rightarrow K^+$	+ 1 electron

Calcium is next. Its structure is 2.8.8.2. It reacts like magnesium 2.8.2 and beryllium 2.2.

The third shell is not filled by eight electrons. Scandium follows calcium. It has one more electron in its atom. This extra electron is not in the fourth shell. The structure of scandium is 2.8.**9**.2.

Calcium
Element 20
Structure 2.8.8.2

Scandium
Element 21
2.8.9.2

The next nine elements fill the third shell. Titanium, element 22, has the structure 2.8.10.2, Vanadium 2.8.11.2, . . . Iron 2.8.14.2. The last one is the metal zinc, 2.8.18.2. These are the transition metals. Most have two electrons in the outside shell. They are metals.

Iron   Fe 2.8.14.2

$$Fe \; 2.8.14.2 \rightarrow Fe^{2+} + 2e$$
2.8.14

Summary

Metals and non-metals differ in properties. Most metals have metallic lustre, high density and high melting point. They all conduct electric current and heat. These are physical properties.

Some metals burn in air. Others only tarnish. The result is an oxide. If it dissolves in water the oxide forms an alkali. Alkalis put hydroxyl, OH^-, ions into solution. The solution turns litmus blue.

Very active metals react with cold water. Alkali forms in solution. Reaction is exothermic. Many heated metals react with steam. The metal oxide forms. Hydrogen is formed in both cases.

All the reactions are examples of oxidation and reduction. $Mg \rightarrow MgO$ is oxidation (oxygen added). It is also $Mg \rightarrow Mg^{2+}$. The atom loses electrons. Oxidation is gaining oxygen and losing electrons. Reduction is losing oxygen or gaining electrons. They happen at the same time in redox reactions.

More active metals can displace less active ones. They are displaced from oxides and other compounds. All these reactions place a metal in the Activity Series. So do its reactions with acids.

Extraction means making a metal from its ores. Very active metals are difficult to extract. They are made by electrolysis of a molten salt. Less active metals are easier to extract. Their oxides are reduced by heating with coke. Very inactive metals occur 'native'. They were used early in history.

Mendeleef listed elements in order of relative atomic mass. He started with hydrogen. The result is called the Periodic Table. Vertical lines are called groups. Each contains a chemical family. Horizontal lines are called periods. They begin with an alkali metal and end with a noble gas.

Later periods contain a Group VIII. It contains transition metals. They include iron and copper. The Periodic Table is explained by the electron structures of the elements.

Questions

1. Sodium, aluminium, iron, tin, copper, gold. These metals are in order of activity. The most active chemically comes first. Read each statement below. Is it true or false? Say why you think so.

   a) Sodium is extracted by electrolysis.
   b) Iron oxide is reduced by heating it with tin.
   c) Gold is likely to occur 'native'.
   d) Tin is made by reducing its oxide with coke.
   e) Aluminium reduces copper oxide to copper.
   f) Iron in copper sulphate solution becomes coated with a layer of copper.

2. Use the equation to answer the questions below.
   zinc oxide + carbon → zinc + carbon monoxide
   $$ZnO + C \rightarrow Zn + CO$$

   a) Which substance is being oxidized?
   b) In this case what does oxidation mean?
   c) Which substance is being reduced?
   d) What is the reducing agent or reductant?
   e) What name is given to this type of reaction?
   f) What substance is manufactured by this reaction?

3. Look at Group I of the Period Table. Find answers to fill the gaps below.

   All the elements are metals except _____. They are called the _____ metals. The outside shell of each atom contains _____ electron(s). They lose this electron to form _____ ions. The sodium ion is written _____. All the metals react with water to form the gas _____. The water then turns litmus _____. It contains an _____. The most reactive metal in the list is _____

Uncle Dick and Auntie Mabel
Learnt the Periodic Table,
To teach Willie, George and also Grace
How the elements fit into place.
Arranging all the elements
Shows chemistry is common sense.

Group I – the alkali metals

We shall study only sodium and its compounds. Some of the work will not be new. Use it as revision.

The other alkali metals are very much like sodium. Those below it are more active than sodium. Activity increases as we pass down the group. The compounds of the group members are like those of sodium.

Sodium chloride is mined as rock salt. It is also present in sea water. The pure substance is made by crystallization. The crystals are cubic. The crystal lattice contains ions. The ions are not free to move. So solid salt is not a conductor.

Sodium chloride melts at about 800°C. The ions can move in the liquid. It can be electrolysed. The anode is a graphite cylinder. It is surrounded by an iron ring. This acts as the cathode of the cell. Molten sodium chloride contains Na^+ and Cl^- ions.

At the anode $2Cl^- \rightarrow Cl_2 + 2e$
Chlorine is liberated. It does not react with the graphite.
At the cathode $2Na^+ + 2e \rightarrow 2Na$
Liquid sodium rises through the molten salt. If it mixes with chlorine they react. Steel gauze keeps them apart.

Sodium is a useful metal. It is used in sodium vapour lamps. These give a bright orange-yellow light. It is used to make sodium cyanide. This is used in extracting gold. Sodium is used as conductor in atomic reactors.

Group I

lithium	Li
sodium	Na
potassium	K
rubidium	Rb
caesium	Cs
francium	Fr

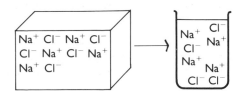

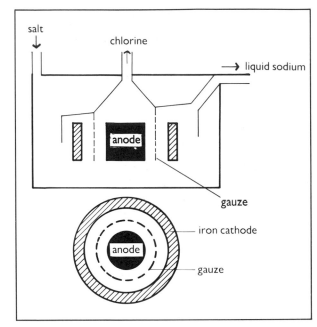

Experiment 18.1 Alkali metals and flame tests

Put sodium compounds on watch glasses. Clean a flame test wire. Dip it in hydrochloric acid. Pick up a few crystals of a compound on the wet end. Hold them in the edge of a blue bunsen flame. Test all the others. Test the chlorides of other alkali metals.

What flame colour do sodium compounds give?
Is the colour caused by the metal in the compound?
List the colours given by other alkali metal salts.

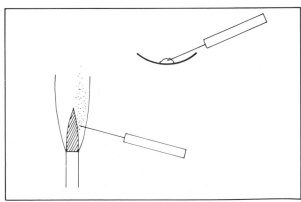

Sodium hydroxide

Experiment 18.2 Properties of sodium hydroxide

1. Put a sodium hydroxide pellet on a watch glass. Let it stand in air for as long as possible.

2. Add four pellets to distilled water in a test tube. Stir the liquid to make them dissolve. Take the temperature before and after dissolving.

3. Make separate solutions of salts. Use sulphates of magnesium, zinc, iron(II), copper(II) and aluminium. Add sodium hydroxide solution to each. Add a little at a time until there is no further change.

4. Warm ammonium chloride crystals with sodium hydroxide solution. Hold a moist litmus paper in the mouth of the tube. Smell the gas with GREAT CARE!

The solid alkali is deliquescent. It takes water vapour from the air. It slowly dissolves in water. When it dissolves heat is produced.

> What are the precipitates formed?
> Which of them dissolve in extra alkali?
> What is the gas formed from ammonium chloride?

We added sodium hydroxide to iron(II) sulphate. Both are ionized in solution. The products must be sodium sulphate and iron(II) hydroxide. All sodium salts dissolve. The green ppt must be iron(II) hydroxide. It slowly turns brown. Why?

$$Fe^{2+} (aq) + 2OH^- (aq) \rightarrow Fe(OH)_2 (s)$$

Aluminium and zinc salts give white ppts. They dissolve if more alkali is added. Copper(II) salts give a blue ppt. This is copper(II) hydroxide.

All ammonium salts react with alkali. Ammonia forms. It has a pungent smell and turns litmus blue.

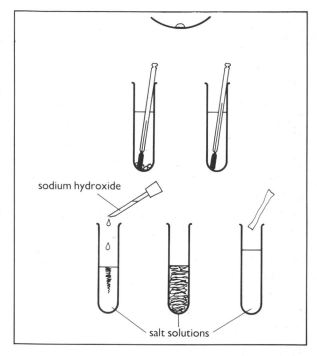

sodium hydroxide

salt solutions

Experiment 18.3 Electrolysis of brine

Use a voltameter as in Experiment 4.2. Add Universal Indicator to brine. Fill the voltameter with it. Collect any gases formed at the electrodes.

> What gas forms at the cathode?
> What happens to the indicator round the cathode?
> What gas forms at the anode?
> What happens to the indicator round the anode?

At the cathode hydrogen appears. The indicator becomes violet. Alkali has been formed. It is mixed with unused brine.
At the anode chlorine appears. It is used up by bleaching the indicator. Some of it dissolves. Its volume is smaller than the volume of hydrogen. There are 3 products, chlorine, hydrogen and sodium hydroxide.

Alkali metal	Colour of flame
lithium	scarlet
sodium	orange-yellow
potassium	lilac

263

Other sodium compounds

Sodium hydroxide is called 'caustic soda'. Sodium carbonate is 'washing soda'. Baking soda is sodium hydrogencarbonate. The last two often give a gas when they react. What gas is it likely to be?

Experiment 18.4 To compare washing soda and baking soda

a) To both solids add dilute hydrochloric acid. Test any gas with lime water. Evaporate one of the solutions. What shape are the crystals formed?

b) Heat a little of each solid in a dry test tube. Test any gas with lime water. Let the tubes cool. Add dilute hydrochloric acid. Again test for a gas.

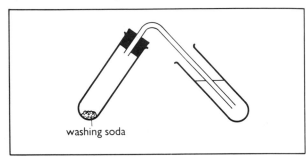
washing soda

Both react with acid. What gas is formed?
What do the cubic crystals show?
When heated, what do both solids give?
Which of them gives carbon dioxide as well?
This heating leaves solids. Both react with acid.
Does this tell us what they are?

Take solutions of each solid. Do these tests.
i) Add drops of Universal Indicator.
ii) Heat each solution. Test any gas. Let the solutions cool. Add dilute acid to each of them.
iii) Take solutions of the sulphates of magnesium, copper and zinc. Add sodium carbonate solution only.

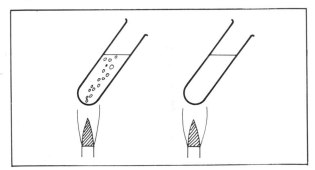

What is the pH of each solution?
Which solution, when heated, gives carbon dioxide?
After heating what does this solution contain?
What are the precipitates formed in test iii)?

Sodium carbonate gives an alkaline solution, pH 10+. With the acid it gives carbon dioxide. The solution gives cubic crystals of salt. So the equation is:

sodium carbonate + the acid → sodium chloride + carbon dioxide + water

$$Na_2CO_3 + 2HCl \rightarrow 2NaCl + CO_2 + H_2O$$

2H from the acid and O from CO_3 give water, H_2O.

Heated carbonate crystals give steam. The solid left reacts with acid. It gives carbon dioxide. It must still be the carbonate. The steam comes from the water of crystallization in Na_2CO_3. $10H_2O$.

The salts in iii) give precipitates. All sodium salts dissolve. The ppts must be metal carbonates.

sodium carbonate + copper sulphate → sodium sulphate + copper carbonate

Sodium hydrogencarbonate is not very soluble. The pH of its solution is about 7. Like the carbonate it reacts with the acid. The products are the same.

$$NaHCO_3 + HCl \rightarrow NaCl + CO_2 + H_2O$$

Gentle heat decomposes the solid. Steam and carbon dioxide are given off. A solid remains. With acid it effervesces. The gas is carbon dioxide. The solid must be sodium carbonate. The word equation:

baking soda heated → sodium carbonate + carbon dioxide + water

Ionic equations are simpler. They show the ions which react. They can show those which do not. For example: Sodium carbonate ions are $2Na^+$ and CO_3^{2-}. The acid ions are H^+ and Cl^-. CO_3^{2-} and H^+ react as shown:
$$CO_3^{2-} + 2H^+ \rightarrow H_2O(aq) + CO_2(g)$$
$2Na^+$ and $2Cl^-$ remain in solution as salt.

Sodium carbonate and copper sulphate solutions give a copper carbonate ppt in sodium sulphate solution.
$$CO_3^{2-}(aq) + Cu^{2+}(aq) \rightarrow CuCO_3(s)$$
$2Na^+$ and SO_4^{2-} remain in solution

Sodium hydrogencarbonate has the ions Na^+ and HCO_3^-. Work out ionic equations for its reactions.

The manufacture of sodium hydroxide

Brine is sodium chloride solution. Experiment 12.6 is electrolysis of brine. Sodium hydroxide forms round the cathode. It is mixed with unused brine. It would be difficult to separate them. The manufacture uses this method. It avoids the separation difficulty.

Brine is fed into the cell. Chlorine appears at the graphite anodes. It leaves through B. The cathode is mercury. Sodium forms there and dissolves in the mercury. The liquid is called sodium amalgam. It runs down the sloping floor.

It passes out of the cell into water. Sodium reacts with water. One product is hydrogen. It leaves through A. The other product is sodium hydroxide solution. It is evaporated. Pure mercury passes back into the cell.

Sodium hydroxide is used in making soap, rayon and paper. It is needed to make pure aluminium oxide. This is used in extracting aluminium. Sodium hydroxide solution reacts with chlorine to form household bleach.

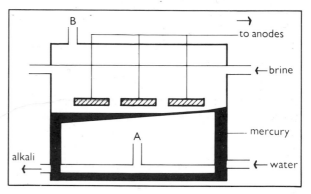

The manufacture of sodium carbonate

This is also made from brine. A concentrated solution is sprayed down a tower. Ammonia is passed in. The solution is sprayed down a second tower. Carbon dioxide is passed into it.

Both towers have bubble caps. They ensure that gases and solution mix well. The gases dissolve. Both saturate the solution. A precipitate forms.

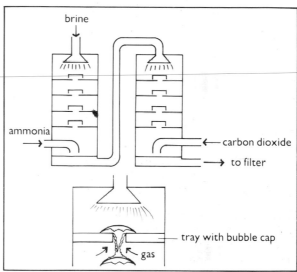

ammonia + water + carbon dioxide → hydrogencarbonate + ammonium ions

$$NH_3 + H_2O + CO_2 \rightarrow HCO_3^- + NH_4^+$$

The ions present in brine are Na^+ and Cl^-.

The ppt is sodium hydrogencarbonate, $NaHCO_3$. It is filtered off. Heating forms the carbonate and carbon dioxide. This is used again.

The filtrate is evaporated. This gives ammonium chloride. Heated with lime it gives back ammonia.

Heating limestone gives lime and carbon dioxide.

Sodium carbonate is used in making soap, rayon, glass and paint. It is also a water softener.

The hydrogencarbonate is baking powder. Mixed with a weak, solid acid it forms baking powder. The soda or powder is mixed with dough. As the dough is heated they react. Both give carbon dioxide. These bubbles in the dough make it porous. It is 'raised'. Baking soda is used in fire extinguishers. It is mixed with a foam producer. They form carbon dioxide. The foam holds it over the flames. The fire is put out.

Group II – the alkaline earth metals

Limestone, chalk and marble are calcium carbonate. Dolomite is a rock containing magnesium carbonate and calcium carbonate. We have used both elements and their compounds. The drawings will remind you of experiments to do again.

beryllium	Be	2.2
magnesium	Mg	2.8.2
calcium	Ca	2.8.2
strontium	Sr	2.8.18.8.2
barium	Ba	2.8.18.18.8.2
radium	Ra	

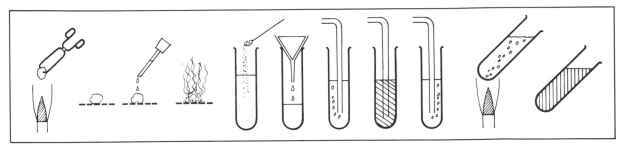

a) Heat chalk or marble. Let it cool. Add drops of cold water. Shake the powder with water. Filter. Test the solution with litmus paper. Pass carbon dioxide through it. Heat the final solution.

b) Burn magnesium in air. Burn calcium. Dip a flame test wire in hydrochloric acid. Do a flame test on each product. Shake some of each product in water. Test the solution with litmus paper.

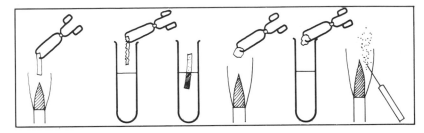

c) Heat magnesium in water. Drop calcium into water. Collect the gas. Test it with a lighted splint. Test the two solutions with litmus. Which is the more active metal? Write equations.

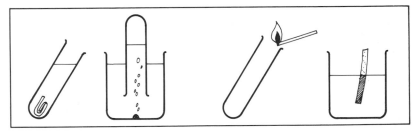

d) Add dilute hydrochloric acid to marble. Collect a gas jar of gas. Put burning magnesium into it. Try to dissolve the product in dilute acid. Which part does not dissolve in the acid? Do a flame test on the solution in the flask.

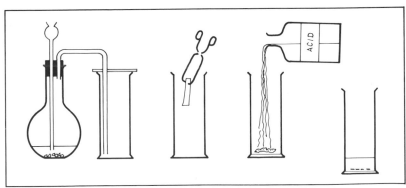

Look at the position of strontium in the Group. Elements in a group are alike. Each metal is more active than those above it.

What compound is the common strontium ore likely to be?
What is the formula of its carbonate?
Will it decompose on heating?
If so, is it harder to decompose than chalk?
Do strontium oxide and water react?
Is strontium hydroxide an alkali?
Does strontium metal react with water?
Will strontium burn in air?
What is the formula of strontium oxide?
Write down the formula of the strontium ion.

Questions on Groups I and II

1. Write this out. Fill in the spaces.

   The apparatus drawn below is used for _____.
   It is called a _____. It has _____ electrodes.
   It contains brine which is sodium _____ solution.
   The Universal Indicator in it is coloured _____.
   This shows that the solution is _____.
   Gases have collected above electrodes A and B.
   The larger volume is above _____. This gas is
   _____.

   This means that electrode A is the _____.
   The indicator round it becomes _____ in colour.
   This colour shows an _____. It is sodium _____.
   It is used in making _____ and _____.
   Electrode B is the _____. The gas set free at B is
   _____.

   It _____ the indicator colourless.

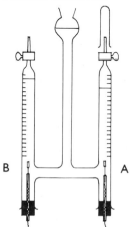

2. What is the chemical name of a) washing soda,
   b) baking soda?
   Both are made by the Solvay Process. Name the two main raw materials used in this method.

What other substance is needed?
How is this other substance got back at the end?

A precipitate of sodium hydrogencarbonate is formed. Is it insoluble in water?
How is it used to make sodium carbonate?
Give two important uses of this carbonate.
Give two uses of baking soda other than baking.

3. What is the everyday name of sodium chloride?
   What is its formula?
   What two ions are there in its crystals?
   If sodium chloride contains ions why is the solid a non-conductor?
   Why is the molten salt a conductor?
   When it is electrolysed, what forms at a) the cathode, b) the anode?
   Electrolysis of sodium chloride solution gives three products. What are they?
   Why does the industrial method use a mercury cathode?

4. Substance A is a shiny white solid. It is strongly heated. It loses mass or weight. It becomes a dull white solid B. Drops of water are added to B when it is cold. It crumbles to a white powder C. Steam is given off. C is shaken with water and filtered. The clear filtrate turns litmus blue. Carbon dioxide is blown through it. It becomes first 'milky' and then clear. Heating the clear liquid makes it 'milky' again.
   These compounds colour the flame brick-red.

   What substance could the shiny white solid A be?
   Explain why A loses mass or weight on heating?
   What is the name of the dull white solid B?
   How do we know that B reacts with water?
   What is the new substance formed, the powder C?
   What kind of substance does litmus show C to be?
   Why does carbon dioxide make the solution of C 'milky'?
   The milky solution becomes clear. Which kind of hardness does it contain?
   What does hard water mean?

Answers to questions on strontium	
carbonate	Yes
$SrCO_3$	Yes
Yes	Yes
harder	SrO
Yes	Sr^{2+}

267

Aluminium – a Group III metal

Aluminium compounds are part of many rocks and clay. The main ore is bauxite. This is aluminium oxide, Al_2O_3. There is plenty of it. Yet finding a way to extract the metal proved to be difficult.

Carbon will reduce the oxide. But it needs a high temperature. At this temperature carbon reacts with aluminium. The carbide forms, not the metal.

Electrolysis could be used. It needs a liquid which contains aluminium ions. Molten aluminium oxide is possible. It melts at about 2000°C. This is too hot for use. The oxide does not dissolve in water.

Molten aluminium chloride? It melts easily. But it is covalent. The liquid contains no aluminium ions. It will not conduct a current.

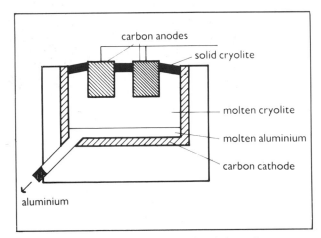

Another aluminium ore is cryolite. It is sodium aluminium fluoride. It is made pure and melted. Pure oxide is made from bauxite. It is dissolved in molten cryolite at 900°C. The cell is lined with carbon. This lining acts as the cathode.

At the cathode, aluminium is set free. At 900°C it is a liquid. It is run out as the pure metal. At the carbon anodes, oxygen forms. It makes the anodes burn away. They have to be replaced.

Aluminium has very useful physical properties. It keeps its shiny lustre. It is an attractive metal. Its density is low for a metal. It is strong. It is a good conductor. It forms alloys with other metals. These are light and strong. Aluminium has many uses.

I   II   III IV V
Na Mg Al Si P
2.8.3

$$Al\ (2.8.3) \longrightarrow Al^{3+} + 3e$$

Experiment 18.5 Chemical properties of aluminium

a) Add aluminium foil to the three common dilute acids. Add foil to sodium hydroxide solution. If nothing happens, warm the mixtures. Test any gases formed.

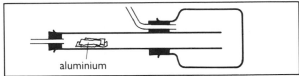

*b) Pass steam over heated aluminium powder as in Experiment 10.3.

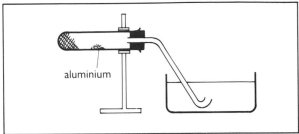

*c) Pass chlorine or hydrogen chloride over heated aluminium foil. Use the apparatus of Experiment 6.14.

d) Hold a strip of aluminium foil in the bunsen flame.

*e) Put a drop of dilute hydrochloric acid on a piece of foil. Add a tiny drop of mercury. Rub the mixture over the foil with a rod. Let it stand.

*f) Repeat the Thermit reaction (page 74).

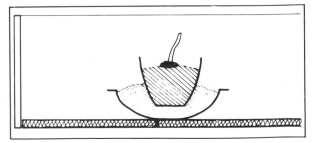

You have used some of these tests before.
Do they show aluminium to be an active metal?
With which substance does it *not react*?

With acids and alkalis the metal gives hydrogen. It reduces steam. It reduces oxides of other metals. Yet it does not easily burn. It appears not to tarnish in air. It does not react with nitric acid.

It reacts with chlorine and hydrogen chloride. Both form the same chloride. It is a covalent solid, formula Al_2Cl_6.
Aluminium is very active. It only appears not to be so.
This has a simple explanation.

Aluminium does tarnish in air. A layer of oxide forms on the surface. This layer is very thin. The shiny surface of the metal shows through it. So the metal appears not to change. The layer of oxide protects the metal.

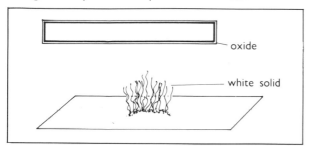

To react, a substance must reach the metal. To do this it must dissolve the oxide layer. This may take time. Reaction is slow to start. It then can be vigorous. If the layer stays, no reaction occurs.

For example, hydrochloric acid reacts with the oxide. Mercury dissolves the exposed metal. Air gets at it. White solid oxide 'grows' on the metal.

Aluminium has many uses from food containers to speed boats.

Experiment 18.6 Aluminium compounds and the Al^{3+} ion

Add aqueous ammonia to aluminium sulphate solution. Divide the mixture into three parts.
a) **To one part add a drop of litmus solution. Shake the mixture. Centrifuge or filter it.**
b) **To the second add dilute acid.**
c) **To the third part add sodium hydroxide solution.**
Add drops of the alkali to aluminium sulphate solution.

Ammonia gives a white jelly-like ppt.
What is it?
What happens to the litmus?

Manufacturing blocks of aluminium.

The sulphate contains aluminium ions. Alkalis give OH^-, hydroxyl ions. The jelly-like ppt is aluminium hydroxide.

$$Al^{3+}(aq) + 3OH^-(aq) \rightarrow Al(OH)_3(s)$$

The ppt reacts with acids. It also reacts with alkalis. This is why sodium hydroxide dissolves it. It acts as an acid with strong alkalis. It acts as a base with acids. Substances like this are called **amphoteric.**

Litmus makes the mixture blue. The filtrate is colourless. The ppt has absorbed the litmus dye. Aluminium oxide does this too. The protective layer on the metal is oxide.

Aluminium and its alloys have an enormous number of uses. They are strong and light. They are used for making cars, aircraft, ships, trains, cooking vessels, bottle tops, cans, building materials and many more.

The oxide layer on the metal can be made thicker. The metal is used as an anode. The cell contains dilute sulphuric acid. Oxygen forms at the anode. Oxygen reacts with aluminium. More oxide forms. The process is called 'anodizing'. Aluminium panels can be coloured. The oxide layer on the metal absorbs dyes.

Aluminium foil is used in packaging. Its lightness and the oxide layer make it ideal 'silver' paper.

269

Iron – a transition element

The transition elements are metals.
They all have the physical properties
of a metal. Can you write these down?

Iron is silvery grey. It has metallic lustre.
Why does it rarely look like this?
Pure iron is fairly hard but not brittle.

Will the bulb light up when we switch on?
What property of iron does this show?

Will the balance show 0.1 g, 1 g or 8 g?
What property of iron does this show?

Which is the most likely melting point of iron:
0°C, 200°C or 1500°C?

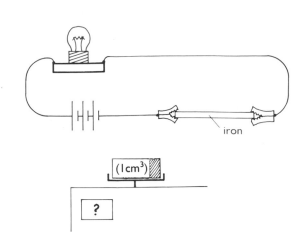

II   VIII   III
Ca Sc Ti V Cr Mn Fe Co Ni Cu Zn Ga
2.8.14.2

Experiment 18.7 The chemical properties of iron

Put a little iron filings in a test tube. Add some dilute
sulphuric acid. Let the mixture stand. Warm it gently if
nothing happens. Test any gas. When the reaction has
almost stopped, filter.

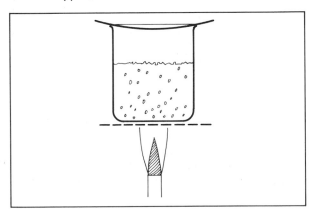

Add a few drops of dilute sulphuric acid to the filtrate.
Evaporate some of it to form crystals. Divide the rest
into four. Test each part with one of the following:
a) aqueous ammonia,
b) sodium hydroxide solution,
c) potassium manganate(VII) solution made acid with
dilute sulphuric acid,
d) hydrogen peroxide.

Add aqueous ammonia to the final solutions in c) and d).

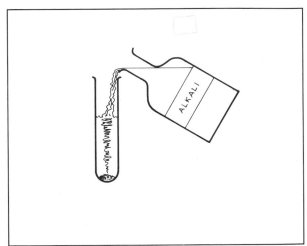

The filter takes out iron. Is the filtrate coloured?
Does the filtrate give coloured crystals?
The filtrate gives a ppt with alkalis. What colour is it?
Potassium manganate(VII) solution oxidizes the filtrate.
How do we know that it does?
Alkali is added after oxidation. What colour is the ppt?

Iron reacts with the acid. The gas formed is hydrogen.
Metal and acid give a salt. The filtrate is pale green. It
gives green crystals. They are iron(II) sulphate,
$FeSO_4.7H_2O$. They contain water of crystallization.

The electron structure of iron is 2.8.14.2. An iron atom loses two electrons from its outside shell. It loses them to hydrogen ions from the acid. The iron(II) ion formed is Fe^{2+}.

$$Fe(s) + 2H^+(aq) \rightarrow Fe^{2+}(aq) + H_2(g)$$

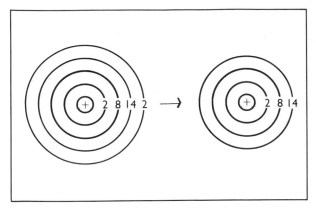

Alkali provides hydroxyl ions, OH^-. They react with Fe^{2+}. The grey green ppt is iron(II) hydroxide.

$$Fe^{2+}(aq) + 2OH^-(aq) \rightarrow Fe(OH)_2(s)$$

Fe^{2+} 2.8.14. does not have noble gas structure. It can lose one more electron to form Fe^{3+}. This is the iron(III) ion. Loss of electrons is oxidation.

$$Fe^{2+} - e \rightarrow Fe^{3+}$$

We know that potassium manganate(VII) oxidizes iron(II). It shows this by becoming colourless. Iron(III) sulphate forms in solution. It also gives a ppt with alkali. The colour of this ppt is rusty brown.

$$Fe^{3+}(aq) + 3OH^-(aq) \rightarrow Fe(OH)_3(s)$$

Iron has two different ions. It has two sets of compounds. Alkali added to a solution shows which one.

Iron(II) compounds give a grey green ppt.

Iron(III) compounds give a brown ppt.

*Experiment 18.8 The two chlorides of iron

Set up the apparatus as shown in a fume cupboard. Put iron wool in the tube. Pass hydrogen chloride through it. Heat the iron gently and then strongly.

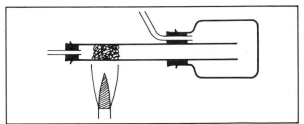

Set up the apparatus a second time. This time pass chlorine over the heated iron. In each case a solid collects in the bottle. Dissolve each separately in water. Take a sample of each solution. Add aqueous ammonia to it.

Take a second sample of each solution. To the one from hydrogen chloride add acidified potassium manganate(VII) solution. To the other add sulphuric acid and zinc. Leave it to stand. After five minutes pour off the solution. Add aqueous ammonia to both liquids.

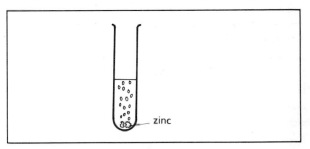

Do the two solids formed appear to be different?
Do they both dissolve in water?
What colour is each of the solutions?
Which of them contains iron(II) chloride?
Does potassium manganate(VII) solution oxidize it to iron(III)?
Chlorine gives iron(III) chloride. How do we know?
What do zinc and acid do to iron(III) chloride?

$Fe + 2HCl \rightarrow H_2 + FeCl_2$ iron(II) chloride
ppt with alkali is grey green

$2Fe + 3Cl_2 \rightarrow 2FeCl_3$ iron(III) chloride
ppt with alkali is rusty brown

Potassium manganate(VII) oxidizes iron(II) to iron(III). Hydrogen from zinc and acid reduces iron(III) to iron(II).

271

Other iron compounds

Experiment 18.9 The three oxides of iron

a) Heat iron(II) oxalate powder. Put a lighted splint at the mouth of the test tube. Pour the hot powder out of the tube. Let it fall from a height on to white paper. Rub the powder over the paper.

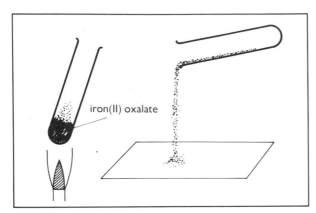

iron(II) oxalate

b) Set up the apparatus. Clean some iron wire. Put it with iron powder in the tube. Heat it so that steam passes over it. Test the gas. Examine the iron.

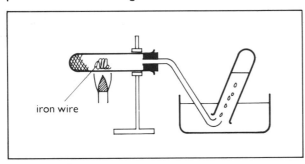

iron wire

What is the colour of iron(II) oxalate?
How do you know that it gives off gas?
What colour is the powder which remains?
What happens as it falls through the air?
What colour is the powder on the paper?
Steam passes over the heated iron in the tube. What is the colour of the solid which remains?

The yellow oxalate 'boils'. It is giving off gas. The gas burns. The black powder is iron(II) oxide, FeO. It falls through cold air. It glows white hot. This heat must come from a reaction. The powder on the paper is brown. It is iron(III) oxide. Iron(II) oxide combines with oxygen from air. $4FeO + O_2 \rightarrow 2Fe_2O_3$

Iron is silvery grey. It reacts with steam to give hydrogen. The substance left is blue-black. It is magnetic iron oxide, Fe_3O_4.

c) Rub iron wire with emery paper. Hold it in the bunsen flame. Drop iron filings through the flame. Catch the product on white paper.

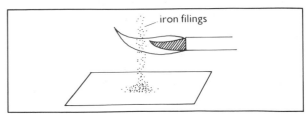

iron filings

Which oxide is it?

Iron has three oxides. Iron(II) oxide is FeO. Air can oxidize it to Fe_2O_3. This is addition of oxygen. It is also loss of electrons by iron(II). $Fe^{2+} - e \rightarrow Fe^{3+}$. Heating iron gives the third oxide, Fe_3O_4. It is blue-black. It behaves like FeO and Fe_2O_3 combined.

Many metals tarnish in air. Iron shows it most. Its tarnish is red-brown rust. Everyday life shows that rusting needs water. What else is needed?

Experiment 18.10 Does rusting use air?

Wet the inside of a test tube. Sprinkle iron filings inside it. Turn it upside down in water. Measure the height of air in the tube. Leave it for a few days. Measure the height of the air again. Take out the tube. Test the air in it with a lighted splint.

What percentage of the air is used up?
What happens to the lighted splint?
What gas is used up during rusting?

These are some results:
first height of air = 14 cm
height after rusting = 11 cm

The volume of air is roughly proportional to height. 3 cm used out of 14 cm is about 21%. The lighted splint goes out. Both show that rusting uses oxygen. Can you guess what iron compound rust might be?

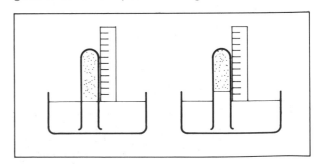

Experiment 18.11 Are other substances needed for rusting?

Clean some nails with emery paper. Set them up in test tubes as shown. Leave them for several days.

1 This tube contains water and air (carbon dioxide and oxygen). It is a control.

2 Boil some water in this tube. Pass carbon dioxide through it for a few minutes. Stopper the tube.

3 This tube contains anhydrous calcium chloride. It takes out water from the air in the tube.

4 This tube contains sodium hydroxide solution. It removes any carbon dioxide.

5 Fill the tube with water. Boil the water for two minutes. Lower the nails into the water. Stopper it with no air inside.

In which tubes do the nails not rust?
Which substances have been taken out of these tubes?
What substances are needed if iron is to rust?

Tube 1 has water and air. Air provides carbon dioxide and oxygen. The nails rust rapidly. Boiling water drives out dissolved air. Therefore:

Tube 2 contains no oxygen.
Tube 3 contains no water.
Tube 4 has no carbon dioxide.
Steam pushed air out of tube 5.

In each tube there is little or no rusting. Each one has one substance missing. So rusting needs water, carbon dioxide and oxygen. With all three it happens. Removing one prevents it. Other substances can replace carbon dioxide. Salt makes rusting more rapid.

Aluminium forms an oxide layer. It is not porous. It clings to and protects the metal. Rust is porous. It flakes off. Substances get at the iron beneath. Rusting goes on. It can be prevented or slowed down. Water and air are kept away.

Iron can be painted. It can be coated with other metals. Zinc, tin, nickel and chromium are all used. The seal must be perfect. Cracks allow water and air to get in. The iron rusts.

Iron cranes and ships all must,
Like any iron, turn to rust.
So paint them all with loving care
To keep out water and the air,
Or oxidation at the dockside
Will turn them into iron oxide.

Summary

Iron has two sets of compounds. This is because it forms two different ions. They are iron(II) Fe^{2+}, and iron(III), Fe^{3+}. Its compounds are coloured. Oxidization is

$$Fe^{2+} - e \rightarrow Fe^{3+}$$

The reverse change is reduction. Both ions in solution give a ppt with alkali. Iron(II) hydroxide is grey green. Iron(III) hydroxide is rusty-brown. Alkali shows which ions are present in solution. Iron has three oxides.

Iron is important. So are its alloys. They are the steels. They rust unless they are protected.

Iron and steel

There are four main ores of iron. Two are oxides. Haematite is iron(III) oxide. Magnetite is Fe_3O_4. There are also sulphide and carbonate ores. These are roasted in air. They give iron(III) oxide.

Iron is extracted from the oxide. It is mixed with coke and limestone. The mixture is fed into the top of a blast furnace. This is a steel tower about 30 metres high. It is lined with firebrick.

The 'blast' is air mixed with oxygen. It is heated to 900 °C and blown into the furnace. The air tubes are called tuyeres (twyers). The coke burns in the oxygen. This reaction is exothermic. The furnace temperature rises to about 1900 °C.

$$C + O_2 \rightarrow CO_2$$

The carbon dioxide rises through the mixture. Coke reacts with it. Carbon monoxide is formed. This reaction is endothermic. The temperature falls.

$$CO_2(g) + C(s) \rightarrow 2CO(g)$$

The monoxide is a gas. It reduces iron(III) oxide to iron. This melts and runs down through the furnace. It is a liquid so substances dissolve in it. Impure liquid iron falls to the bottom.

$$Fe_2O_3(s) + 3CO(g) \rightarrow 2Fe(l) + 3CO_2(g)$$

The high temperature decomposes limestone. Calcium oxide, quicklime, forms.

$$CaCO_3 \rightarrow CaO(s) + CO_2(g)$$

Quicklime reacts with impurities in the iron ore. For example, sand forms calcium silicate. The result is liquid slag. This runs down the furnace. Slag floats on the molten iron. Both slag and iron are tapped off.

Mixture fed in at the top keeps the furnace full. The process need only stop for repairs. Waste gas leaves at the top. It is piped off and burnt. The heat is used to heat the blast. The slag is used in the building industry.

The impure iron contains up to 4% carbon. It may contain other elements such as phosphorus. These make it harder but more brittle. They are removed from the iron before turning it into steel.

A blast furnace.

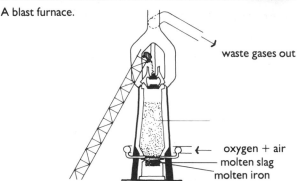

waste gases out

oxygen + air
molten slag
molten iron

Tapping iron from a furnace.

There are two main methods of doing this. Both use the molten metal from the furnace. Some impurities are burnt out. Some react with basic oxides such as quicklime.

1. The Open Hearth method uses an open hearth furnace which is a shallow basin. The furnace is lined with fire brick. The molten iron is put in. Ore and scrap iron are added. Lime is the basic oxide used. The total charge may be 500 tonnes.

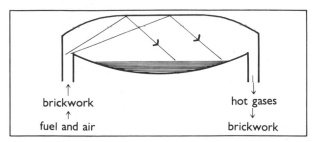

Liquid and gas fuel are burnt at one end. The flame is deflected by the roof. It heats the charge. Carbon is burnt out of the iron. Other impurities react with the lime. The slag floats on the iron.

Hot gases pass out through brickwork. They heat it. The flow of fuel and air is reversed. It is made much hotter by the brickwork. The iron becomes pure.

2. The Basic Oxygen method uses a converter.

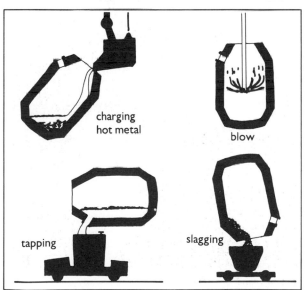

Pouring hot metal into a basic oxygen furnace.

This is a steel vessel lined with fire brick. It holds about 300 tonnes of molten iron. Scrap iron and lime are added. Oxygen under pressure is blown in. It passes through a water cooled tube called a lance.

Lime removes impurites. Carbon is burnt out. A slag forms. It floats on the pure iron. In both methods this iron is turned into steel. Other elements are dissolved in it to give alloys. The amounts are decided by the properties needed.

Ordinary steels are alloys of carbon and iron. The carbon content varies from 0.2 to 1.5%. Special steels contain other metals. Steel for magnets will contain cobalt. Stainless steel contains nickel and chromium. They are examples of the fourteen million tonnes of steel manufactured in Britain every year.

A third method of making steel is the electric arc method. The furnace is charged with ore, scrap iron and lime. Heat comes from an electric arc. Pure steel is formed. It has the correct amount of carbon. The steel is poured out as a liquid.

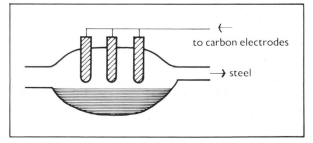

Questions

In each question five answers are given. They are labelled A, B, C, D and E. Only one answer is correct. Choose this one. Write its letter on your answer sheet.

1. Water is
    A   an element.
    B   a compound of carbon and oxygen.
    C   a mixture of hydrogen and oxygen.
    D   a compound of hydrogen and oxygen.
    E   a compound of hydrogen and carbon.

2. Both diamond and graphite
    A   conduct an electric current.
    B   form carbon dioxide when they burn.
    C   are black and crystalline.
    D   have the same density.
    E   are used to drill through rock.

3. Oxygen is made in industry by
    A   heating black copper oxide.
    B   photosynthesis.
    C   heating red mercury oxide.
    D   using hydrogen peroxide solution.
    E   boiling liquid air.

4. Which of these changes is a chemical one?
    A   dissolving salt in water.
    B   changing rhombic to monoclinic sulphur.
    C   melting ice to water.
    D   burning petrol in a car engine.
    E   evaporating salt solution.

5. In pure water, Universal Indicator is A red, B orange, C green, D blue, E violet.

6. In alkali, Universal Indicator is coloured A red, B orange, C yellow, D green, E blue.

7. A good conductor of electric current is
    A   hydrogen chloride
    B   a solution of hydrogen chloride in water
    C   a solution of hydrogen chloride in methylbenzene
    D   pure water
    E   methylbenzene.

8. A sodium ion differs from a sodium atom in having
    A   one less proton
    B   one less electron
    C   one more proton
    D   one more electron
    E   one more neutron.

9. An aqueous solution of copper sulphate is electrolysed. Both electrodes are platinum. The products could be

	at the positive anode	at the negative cathode
A	copper	sulphate
B	hydrogen	copper
C	oxygen	copper
D	hydrogen	oxygen
E	oxygen	hydrogen

Questions 10 to 15 are about five methods used in Chemistry. They are: A filtration, B distillation, C chromatography, D crystallization, E using a magnet. Which of the five is used to:

10. get pure water from a salt solution

11. make salt crystals from salt solution

12. separate the coloured substances in flower petals

13. get iron from a mixture of iron and tin

14. get clear water from muddy water

15. take unused copper oxide from copper sulphate solution

Questions 16 to 21 are about five kinds of chemical change. They are: A reduction, B combination, C decomposition, D neutralization, E forming a precipitate. Which one describes each change shown below?

16. tin oxide + carbon → carbon dioxide + tin

17. magnesium burning in oxygen

18. acid + alkali → salt + water

19. mercury oxide heated → mercury + oxygen

20. silver nitrate + sodium chloride (both in solution in water)

21. hydrogen peroxide → oxygen + water

22. A certain metal: (i) does not react with dilute sulphuric acid to give hydrogen, (ii) becomes coated with copper when put into copper sulphate solution In the Activity Series this metal would be placed
    A   above sodium
    B   below sodium and above magnesium
    C   below iron but above hydrogen
    D   below hydrogen but above copper
    E   below copper

23. Another metal: (i) does not react with cold water but does react with steam. Hydrogen is formed. (ii) can be used to reduce iron oxide to iron. In the Activity Series this metal would be placed
    A   above sodium
    B   below sodium but above iron
    C   below iron but above hydrogen
    D   below hydrogen but above copper
    E   below copper

In Questions 24 to 28 use C = 12, O = 16, Ca = 40.

24. C = 12 shows
    A   the atomic number of carbon
    B   the formula mass of carbon
    C   the relative atomic mass of calcium
    D   the formula mass of calcium
    E   the relative atomic mass of carbon

25. The mass of a calcium atom is
    A   three times the mass of oxygen atom
    B   2.5 times the mass of an oxygen atom
    C   twice the mass of an oxygen atom
    D   equal to the mass of an oxygen atom
    E   less than the mass of an oxygen atom

26. The formula mass of carbon dioxide, CO_2, is:
    A 12, B 28, C 44, D 56, E 60.

27. One mole of chalk, $CaCO_3$, is:
    A 68, B 124, C 100, D 100 grams, E 116.

28. Heat decomposes chalk: $CaCO_3 \rightarrow CaO + CO_2$
    One mole of chalk gives
    A   one mole of carbon dioxide
    B   28 grams of carbon dioxide
    C   2 moles of carbon dioxide
    D   3 moles of carbon dioxide
    E   one mole of calcium dioxide

Questions 29 to 32: the apparatus is shown in the drawing. We used it to find the volume of oxygen in the atmosphere. Choose the best answer.

29. The two pieces of apparatus labelled X are
    A   test tubes
    B   measuring cylinders
    C   glass tubing
    D   syringes
    E   condensers

30. The metal in the tube is: A magnesium, B copper, C mercury, D iron, E tin.

31. This metal takes out of the air: A nitrogen, B carbon dioxide, C water vapour, D oxygen, E some other gas.

32. At the start the apparatus contains 50 cm^3 of air. The metal removes all possible gas. The volume left at the end is: A 40 cm^3, B 30 cm^3, C 20 cm^3, D 10 cm^3, E some other volume.

Questions 33 to 38: elements may be placed in the following groups:
A Alkali metals, B Noble gases, C Group II metals, D Halogens, E Transition metals. In which group would you place an element:

33. whose molecule contains two atoms?
34. whose atom has the electron configuration 2.8.8?
35. whose atom easily loses one electron?
36. whose ions give coloured solutions?
37. whose atom forms two different positive ions?
38. whose atom forms the ion, X^{2+}, only?

Questions 39 to 42: the following are sets of three elements: A (Cl, Br, I), B (He, Ne, Ar), C (Li, Na, K), D (Na, Mg, Al), E (Be, Mg, Ca).
Which set

39. forms few if any compounds?
40. has the electron configurations 2.8.1, 2.8.2, 2.8.3?
41. contains alkali metals?
42. contains halogens?
43. contains elements from more than one group of the Periodic Table?
44. has atoms with an outside shell of 7 electrons?

Questions 45 and 46: electrical conductivity is the result of movement of particles. Which of the particles below move when:

45. a metal conducts a current?
46. a salt solution conducts?
    A atoms, B electrons, C protons, D ions, E neutrons.

Questions 47 to 52: five terms used in electrolysis are:
A electrolyte, B cathode, C anode, D anion, E cation.
Select the correct one for:

47. the liquid through which the current passes,
48. a negatively charged ion,
49. the positive electrode,
50. the electrode at which metals and hydrogen appear,
51. a positively charged ion,
52. the electrode at which oxygen often appears.

Questions 53 to 71: each question has four statements. One or more of them is correct. Decide which are the correct ones. Then write down:

A if only 1, 2 and 3 are correct,
B if only 1 and 3 are correct,
C if only 2 and 4 are correct,
D if only 4 is correct,
E if another statement or group of statments is correct.

53. The gas hydrogen
   1 is less dense than air
   2 can reduce some metal oxides
   3 forms an explosive mixture with air
   4 has a choking smell

54. Acids can be neutralized by
   1 sodium hydroxide
   2 sodium chloride
   3 sodium carbonate
   4 sodium sulphate

55. The atmosphere contains
   1 about 21% nitrogen
   2 more carbon dioxide than oxygen
   3 more nitrogen than oxygen
   4 about 1% hydrogen

56. The reaction of an acid with an alkali produces
   1 an element
   2 a salt
   3 hydrogen
   4 water

57. Substances which lose mass on heating include
   1 copper carbonate
   2 black copper oxide
   3 copper sulphate crystals
   4 copper metal

58. Calcium oxide is also called
   1 limestone
   2 slaked lime
   3 chalk
   4 quicklime

59. The rate of reaction between calcium carbonate and dilute hydrochloric acid is increased by
   1 using powdered chalk instead of lumps
   2 raising the temperature
   3 increasing the concentration of the acid
   4 using a larger flask to hold the mixture.

60. The nucleus of a large atom contains
   1 electrons only
   2 protons only
   3 neutrons only
   4 protons and neutrons.

61. The element rubidium, Rb, comes below sodium and potassium in Group I of the Periodic Table. You would expect it to
   1 react vigorously with water
   2 have a very high density
   3 to form the ion Rb^+
   4 not tarnish in air.

62. Oxidation is
   1 the loss of oxygen by a substance
   2 the gain of oxygen by a substance
   3 the gain of electrons
   4 the loss of electrons.

63. The change from a metal atom to its ion, $Mg \rightarrow Mg^{2+}$, is an example of
   1 reduction
   2 oxidation
   3 extraction
   4 ionization.

64. Which statements about nitrogen are true?
   1 its molecule contains two atoms
   2 it burns readily in air
   3 it reacts with hydrogen to form ammonia
   4 it is 21% of the air around us.

65. The gas methane
   1 has the formula CH_4
   2 burns in air forming carbon dioxide and water.
   3 is the chief compound in 'natural' gas
   4 has two isomers.

66. Which gases have a relative molecular mass of 28 if $H = 1, C = 12, N = 14, O = 16$?
   1 carbon monoxide, CO
   2 nitrogen
   3 ethene, C_2H_4
   4 oxygen.

67. Acids
   1 can be neutralized by bases
   2 in water have a pH between 0 and 6
   3 contain hydrogen which gives H^+ ions in solution
   4 react with non-metals to give hydrogen.

68. Two different substances have the formula C_4H_{10}.
    The two substances are called
    1 isotopes
    2 polymers
    3 carbohydrates
    4 isomers.

69. A metal is below aluminium in the Activity Series. It is
    above iron. It would be expected to
    1 react with cold water
    2 react when heated in steam
    3 reduce aluminium oxide to the metal
    4 displace hydrogen from dilute acids.

70. Ethanol
    1 has the formula C_2H_5OH
    2 reacts with acids to form esters
    3 decomposes to form ethene and water
    4 puts hydroxyl ions, OH^-, into solution in water.

71. Polymers are made by
    1 small molecules of the same substance reacting
    2 large molecules decomposing
    3 monomers reacting to form a giant molecule
    4 fermentation of sugar solution.

Questions 72 – 73: Answer each part of these longer
questions.

72. Describe how you could set up a simple cell. Use the
    metals copper and zinc and any solutions you need.
    Which electrode is negative? How would you
    measure the voltage the cell gives? Why does the
    voltage fall away during use? How does a dry cell
    prevent this effect?

73. The apparatus shown contains 20 g of marble chips.

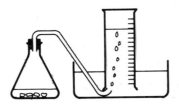

50 cm³ of M hydrochloric acid is added. The
temperature of the acid is 16°C. The gas given off is
collected. The volumes of gas are shown:
volume collected after one minute is 40 cm³
volume collected after two minutes is 70 cm³
a) What volume of gas is collected in the second
   minute?
b) Why is it less than the volume collected in the
   first minute?
c) Would it be even less in the third minute?
The experiment is repeated with 20 g of smaller
lumps of marble. 50 cm³ of M acid at 16°C is used.
d) Will the gas collected in one minute be greater or
   less than 40 cm³? Explain why.
The experiment is repeated with the acid at 32°C.
e) Will the gas collected in one minute be greater or
   less than 40 cm³? Say what you think the volume
   will be?
The experiment is repeated with 50 cm³ of 0.5 M
acid.
f) Will the volume be less or more in the first
   minute? Say what you think the volume will be.

More questions

1. Here is a list of elements. They are in order of chemical activity. The most active element comes first.

calcium, aluminium, iron, lead, hydrogen, copper, gold

Use the list to answer these questions:
a) Which element in the list is not a metal?
b) Which element combines best with oxygen?
c) What is likely to happen when aluminium is heated with lead oxide? Write a word equation for any change.
d) Which elements will not liberate hydrogen from dilute acid?
e) Which metal in the list is most likely to be found 'native'?
f) Will iron displace copper from copper sulphate solution?
g) Which metals have oxides which can be reduced by hydrogen?

2. Three towns, A, B and C, have different water supplies. 50 cm³ samples of each are taken. Soap solution is added until a lasting lather forms. 50 cm³ samples of each are boiled. Soap solution is added. Each is passed through a water softener. Soap solution is added to 50 cm³ of this water. In each case enough is added to give a lasting lather. The results are given in the table below in cm³ of soap solution needed:

	Town A	Town B	Town C
Tap water needs	20	15	10
Boiled tap water needs	16	15	1
Softened water needs	1	1	1

Which town has the hardest water?
Which types of hardness does each town supply contain? What substance causes hardness in water from town B? What substance causes the hardness of town C? How much of this soap solution would distilled water need? Rain in town A falls into a beaker. How much soap would 50 cm³ of it need to give a lasting lather?

3. Salts can be made by several methods. Here is a list of salts: zinc sulphate, copper sulphate, silver chloride, sodium chloride, magnesium carbonate. From it, choose:
a) one which can be made by the action of acid on metal
b) one made by the action of an acid on an oxide
c) one made by precipitation
d) one made by mixing acid and alkali.
Describe in detail how you would make crystals of one of them.

4. Substance A is a white solid. It is heated strongly in a bunsen flame for a long time. It colours the flame brick red. A white solid B is formed. Its mass is less than the mass of A. With water, B crumbles to a powder C. Steam is formed. C is shaken with water. The mixture is filtered. Carbon dioxide is passed into the clear filtrate. It becomes cloudy.
a) Name the substance A.
b) Explain why it loses mass on heating.
c) What does the flame colour mean?
d) Name the substance B.
e) What tells you that B reacts with water?
f) Name the substance C.
g) The filtrate is a solution of C. Why does it become cloudy?
h) What effect would the filtrate have on litmus?
j) Give one everyday use of B and one of C.

5. Hydrogen is burnt in air. The products are cooled. A clear, colourless liquid is formed.

What is its chemical name?
What is its everyday name?

The liquid turns to a solid at 0 °C. What will be its boiling point? Salt is dissolved in the liquid. How will its boiling point alter? How would you get salt crystals from the solution? What will be the shape of the crystals? Describe briefly how you would make pure water from tap water.

6. These are relative atomic masses: Cu = 64, O = 16, H = 1, S = 32.
Which atom has 64 times the mass of a hydrogen atom? 64 g of copper form 80 g of black copper oxide. Work out the formula of the oxide. What is one mole of it? The formula of sulphuric acid is H_2SO_4. What is its relative molecular mass? What is the mass of one mole of it? Write an equation for the reaction between the copper oxide and sulphuric acid. How much acid is needed to react with 8 g of the oxide? How much copper sulphate will be formed?

7. Write one sentence, in each case, to explain what is meant by:
an allotrope; reduction; an element; a chemical change; an atom; a molecule; a mole.

8. Describe an experiment to determine the heat of reaction of an acid with a metal. How can the results of such experiments be used to put metals in order of activity? What is this order called?

Word games

Word changing

Example: Change sand to salt in three moves.
Change one letter at each move to make a new word.
Alchemy! change lead into gold in three moves.
LEAD

. . . .

. . . .

GOLD
Change salt into lime in four moves.
SALT

. . . .

. . . .

. . . .

LIME
Make up your own word changes. COPPER into CARBON?
BOIL into MELT?

SAND
SAN<u>E</u>
SA<u>L</u>E
SAL<u>T</u>

Symbol chain

Write down any symbol such as Pb. Use its second letter to start a new symbol: Pb . . Ba. Now use this second letter for the next: Pb . . Ba . . Al . . Li . . How long can you make the chain?

Anagrams

Arrange the letters of each phrase to make one word. The meaning of the word you are to make is given. For example:

P.C. ROPE	A metal used as a conductor in wire form	COPPER
TRIFLE	To separate a solid from a liquid	_____
PARE A VOTE	To get rid of a liquid by heating it	_____
SEND ONCE	To turn steam back into water	_____
LAST CRY	Lump of solid with a regular shape	_____
BIG RUIN	The scientific name for it is combustion	_____
CURE ED	To take oxygen from a substance	_____
BE RAKE	Vessel for holding liquids	_____
MEET THE MORE	It measures temperature	_____
GO DRY HEN	Gives squeaky pop with lighted splint	_____
LANE RUT	Neither acid nor alkaline	_____

Chemist's alphabet

The smallest bit of an element is an	A _____
Effervescence means	B _____
Dyes in ink are separated by	C _____
One allotrope of carbon is	D _____
A substance which cannot be split up into others is an	E _____
The liquid which runs through a filter paper is the	F _____
A yellow, noble metal is	G _____
The chemical name of water is	H _____
The element which gives violet vapour on heating is	I _____
If you think this alphabet is easy you must be	J _____
	_____
The symbol for the element potassium is	
To find out if a liquid is an acid use	L _____
L atoms or molecules of a substance is called a	M _____
78% of the air we breathe is	N _____
21% of the air we breathe is	O _____
Green leaves of plants make starch by	P _____
There are two limes. Calcium oxide is	Q _____
Taking oxygen from a substance is called	R _____
A yellow non-metal is	S _____
The simplest piece of apparatus is a	T _____
To find out how strong an acid is use	U _____
In strong alkali Universal Indicator is	V _____
Hydrogen oxide is	W _____
One of the noble gases is called	X _____
The colour of gold and sulphur is	Y _____
A more active metal than copper is	Z _____

Now invent an alphabet of your own.

Words

Put one letter in front of each four-letter word to make a five-letter word. The
letters you use must make the name, in the right order, of a well-known metal.

___rack

___bout   Add a letter to each three-letter word to make a four-letter word.

___earn   The added letters make the name of a metal.

___lean   ___top

___deal   ___van

___sing   ___air

___over   ___ale

                ___den

                ___oar

Crosswords

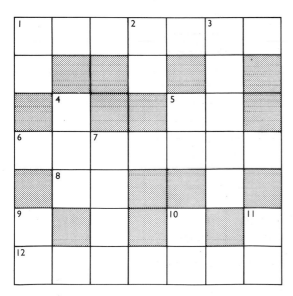

Across

1. A non-metal, often black. (6)
5. A mineral found in the earth. (3)
6. Fine powder. (4)
8. The symbol for lead. (2)
9. The symbol for nickel. (2)
11. To take oxygen from a substance. (6)

Down

1. Pink metal used as a conductor. (6)
2. Symbol of the noble gas argon. (2)
3. Colour of litmus in acid. (3)
4. Metals found in the earth are this. (6)
7. The symbol for the metal tin. (2)
10. The symbol for the metal copper. (2)

Across

1. A piece of solid with a regular shape. (7)
5. The symbol for the metal silver. (2)
6. The element with the symbol Si. (7)
8. The symbol for the metal nickel. (2)
12. A single simple substance. (7)

Down

1. The symbol for the metal calcium. (2)
2. The symbol for the metal tin. (2)
3. A noble gas which is 1% of air. (5)
4. A silvery metal. (3)
5. Guess the symbol of actinium! (2)
7. Calcium oxide or calcium hydroxide. (4)
9. Symbol for the metal iron. (2)
10. Symbol for the noble gas neon. (2)
11. Symbol for the metal platinum. (2)

Glossary

The glossary is a list of important words or terms. The first number refers to the first page on which it is mentioned. Other page numbers refer to the important uses.

Acid: A substance which forms hydrogen ions, H^+, in water/ 44, 46, 66, 190

Activity Series: The metals listed in order of how well they react/ 70, 196, 254

Addition reaction: Two or more substances react to form one product/ 212

Alkali: A substance which puts hydroxyl ions OH^-, into water/ 44, 46, 67, 190

Alkali metal: A metal in Group I of the Periodic Table/ 258, 262

Alkaline earth metal: A metal in Group II of the Periodic Table/ 258, 266

Allotropes: Different forms of the same element/ 130, 134

Allotropy: The existence of allotropes of an element/ 130, 131, 134, 202

Analysis: Tests to find out what a substance contains/ 43

Anion: A negative ion. It moves to the anode during electrolysis/ 154

Anode: A positively charged electrode/ 150

Atom: The smallest particle (bit) of an element/ 110

Atomic number: The number of protons in the nucleus of an atom/ 159

Atomic structure: The way smaller particles are arranged in an atom/ 159

Atomic Theory: Using atoms to explain the facts of science/ 116

Avogadro's Number (L): 602 300 000 000 000 000 000 000 atoms, molecules etc./ 119

α-rays (alpha radiation): A stream of positively-charged helium ions/ 162

β-rays (beta radiation): A stream of electrons/ 162

Boiling point: The temperature at which a liquid boils/ 20, 66

Carbon chemistry: The branch of chemistry dealing with carbon compounds/ 202

Carbon cycle: Changes which control oxygen/carbon dioxide proportions in air/ 52

Catalyst: A substance which alters the rate of reaction. It is unchanged at the end./ 43, 106, 139

Cathode: A negatively-charged electrode/ 150

Cation: A positive ion attracted to the cathode/ 150

Cell: A device for turning chemical energy into electrical energy/ 198

Change of state: Change from solid to liquid or liquid to gas or the reverse/ 106

Chemical change: A change in which at least one new substance is formed/ 33, 66, 106

Chemical properties: The properties of a substance which can make a new substance/ 106

Chemical reaction: Another name for a chemical change/ 33, 66, 106

Chromatography: Separating substances by washing them through a solid/ 24, 66

Classification: Grouping things of the same kind together/ 74, 106

Compound: Two or more elements joined up into one substance/ 39, 74, 106

Condensation reaction: A reaction between two molecules forming a bigger one and a smaller one such as water/ 218, 220

Conductor: A substance which allows an electric current to pass through it/ 150

Corrosion: Loss of metallic lustre by a metal reacting in air/ 80

Covalent bond: A shared pair of electrons joining two atoms/ 186

Covalent compound: A compound held together by covalent bonds/ 186, 188

Crystals: Pieces of a substance which have the same regular shape/ 14, 66

Decompose:	To break down a substance into other substances/ 39, 66, 82
Deliquescent:	Able to take water vapour from air to form a solution
Dehydration:	The removal of water from a substance. It is also the removal of hydrogen and oxygen in the same proportions as in water.
Displacement reaction:	Another name for replacement reaction/ 196
Distillation:	Boiling a liquid so that the vapour formed can be condensed/ 22, 66
Distilled water:	Pure water made by distillation/ 22, 96
Double bond:	Two covalent bonds joining the same atoms/ 207, 212
Effervescence:	The escape of bubbles of gas from a liquid/ 88
Electrode:	The conductor through which a current enters or leaves/ 150
Electrolysis:	Current passing through a liquid forming new substances at the electrodes/ 150
Electrolyte:	The liquid used in electrolysis/ 150
Electron:	Negatively-charged particle, with a mass about $\frac{1}{2000}$ of an H atom/ 158, 180
Electron shell:	Group of electrons round the nucleus of an atom/ 160, 180
Element:	A single substance which cannot be split up into sinpler substances/ 39, 74, 106
e.m.f.:	The electric pressure (in volts) which causes current to pass/ 200
Endothermic reaction:	A reaction in which heat is taken in as the substances react/ 192
Energy cycle:	Another name for the carbon cycle/ 52
Exothermic reaction:	A reaction during which heat is given out/ 192
Filtration:	Separating a solid from a liquid (the **filtrate**) using a filter/ 14
Flame colouration:	The colour which a metal compound gives to a flame/ 43
Formula:	A way of indicating the atoms present in a molecule/ 119
Formula mass:	Relative molecular mass as shown by the formula/ 120
Fuel:	A substance burnt to give heat and energy/ 50, 66
γ-rays (gamma radiation):	A stream of rays similar to X-rays/ 162
Half-life:	The time taken for half of a radioactive element to disintegrate/ 163
Heat of reaction (ΔH):	The heat change when moles react as shown in the equation/ 192
Heat of combustion:	Heat of reaction for combustion/ 193
Heat of displacement:	Heat of reaction for displacement/ 196
Heat of neutralization:	Heat of reaction for neutralization/ 194
Hard water:	Water with dissolved compounds in it which affect soap/ 94
Hydrogen ion:	The ion H^+ which makes a solution acid/ 190
Hydroxyl ion:	The ion OH^- which makes a solution alkaline/ 190
Indicator (acid-base):	A substance with different colours in acid and alkali/ 44, 46, 66
Insulator:	A substance which does not conduct a current/ 69
Ion:	A charged atom or group of atoms/ 150
Ionic bond:	Attraction between ions of opposite charges/ 182
Ionic compound:	A compound held together by ionic bonds/ 182, 188
Isomers:	Substances with the same molecular formulae but different structural formulae/ 205, 210
Isotopes:	Atoms of the same element with different masses
Joule:	The unit of energy, 1000 joules = 1 **kilojoule**/ 192

Mass:	The amount of substance in a substance or object/ 35, 125
Mass number:	The number of protons and neutrons in the nucleus of an atom/ 159
Melting point:	The temperature at which a solid melts/ 20, 66
Metallic lustre:	The surface shine which all metals have/ 32
Mineral:	A substance found in the earth/ 18
Molar or M solution:	1 mole of solute in 1000 cm³ of solvent/ 121
Mole:	Avogadro's number (L) of atoms, molecules, ions or electrons/ 120
Molecule:	A particle which contains two or more atoms/ 110
Monomer:	A small unit. Many polymerize to form a polymer/ 218
Neutral:	Neither acid nor alkaline/ 45, 46, 66
Neutron:	Particle of no charge and mass of 1 atomic unit/ 159
Nuclear reaction:	A change in the nucleus of an atom/ 163
Nucleus:	The tiny centre of an atom/ 158
Oxidation:	Addition of oxygen to a substance or loss of electrons/ 53, 138, 171, 255
Periodic Table:	The elements arranged in order of atomic number/ 258
Permanent hardness:	Hardness not removed by boiling water/ 94
pH:	A number showing the concentration of hydrogen ions/ 190
pH range:	The numbers from 0 to 14 which measure pH/ 190
Photosynthesis:	The process of making sugar and starch in green plants in light/ 54
Physical change:	A change in which no new substance is formed/ 34, 66, 106
Physical properties:	Properties not affecting chemical change/ 106
Plastic:	A substance, usually a polymer, which can be moulded/ 218, 224
Polymer:	The substance formed by small molecules combining/ 218, 224
Precipitate (ppt):	An insoluble solid formed in a liquid/ 147
Primary cell:	An electrical cell which cannot be recharged/ 200
Products of combustion:	The substances formed by burning/ 49
Properties:	The apperance and behaviour of a substance/ 20, 66, 106
Radioactivity:	Radiation from the breakdown of atomic nuclei/ 158, 162
Radioisotope:	A radioactive isotope of an element/ 162
Rate of reaction:	The speed of a chemical change/ 248
Redox reaction:	A reaction in which reduction and oxidation happen/ 255
Reduction:	The loss of oxygen from a substance or the gain of electrons/ 66, 75, 139, 255
Relative atomic mass:	The numbers which compare the masses of atoms of the elements. They are written H = 1, O = 16, etc/ 118
Relative molecular mass:	Mass of a molecule on the relative atomic mass scale/ 120
Replacement reaction:	A change in which one element replaces another/ 77
Reversible reaction:	One in which products react to give the reacting substance/ 101
Saturated compound:	A compound which only reacts by substitution/ 203
Saturated solution:	A solution in which no more solute will dissolve/ 16
Secondary cell:	A cell which can be recharged/ 200
Sewage:	Used water carrying waste matter/ 102
Soft water:	Water which lathers at once when soap is added/ 94
Soluble:	Able to dissolve/ 17, 66
Solute:	The substance which dissolves/ 16, 66

Solution:	A liquid with a substance dissolved in it/ 15, 66
Solvent:	The liquid in which a substance dissolves/ 16, 66
States of matter:	Matter which can be solid, liquid or gas. These are the three states of matter./ 8
Structural formula:	A formula which shows how atoms are joined in an atom/ 204
Substance:	One particular kind of matter/ 8, 66
Substitution reaction:	Replacement of atoms in a molecule by different ones/ 203
Symbol:	The letter or letters which stand for one atom of an element/ 116
Temporary hardness:	Hardness which can be removed by boiling water/ 100
Thermit reaction:	The reduction of a metal oxide by aluminium/ 76
Transition temperature:	The temperature at which one allotrope changes to another allotrope/ 132
Unsaturated compound:	A compound which reacts by addition/ 213
Weight:	The pull of the earth on an object/ 35
Word equation:	A summary of a reaction, naming reactants and products

List of elements

Element	Symbol	Atomic number	Relative atomic mass
Aluminium	Al	13	27
Argon	Ar	18	40
Barium	Ba	56	137
Bromine	Br	35	80
Calcium	Ca	20	40
Carbon	C	6	12
Chlorine	Cl	17	35.5
Chromium	Cr	24	52
Cobalt	Co	27	59
Copper	Cu	29	64
Fluorine	F	9	19
Gold	Au	79	197
Helium	He	2	4
Hydrogen	H	1	1
Iodine	I	53	127
Iron	Fe	26	56
Lead	Pb	82	207
Lithium	Li	3	7

Element	Symbol	Atomic number	Relative atomic mass
Magnesium	Mg	12	24
Manganese	Mn	25	55
Mercury	Hg	80	201
Neon	Ne	10	20
Nickel	Ni	28	59
Nitrogen	N	7	14
Oxygen	O	8	16
Phosphorus	P	15	31
Platinum	Pt	78	195
Potassium	K	19	39
Silicon	Si	14	28
Silver	Ag	47	108
Sodium	Na	11	23
Strontium	Sr	38	88
Sulphur	S	16	32
Tin	Sn	50	119
Uranium	U	92	238
Zinc	Zn	30	65

Periodic Table

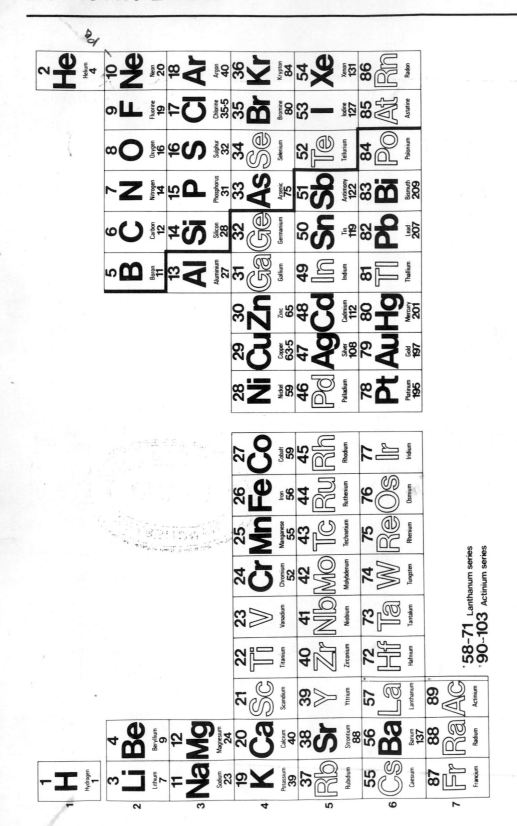

The periodic table shows:

Row 1: 1 H Hydrogen 1 ... 2 He Helium 4

Row 2: 3 Li Lithium 7, 4 Be Beryllium 9 ... 5 B Boron 11, 6 C Carbon 12, 7 N Nitrogen 14, 8 O Oxygen 16, 9 F Fluorine 19, 10 Ne Neon 20

Row 3: 11 Na Sodium 23, 12 Mg Magnesium 24 ... 13 Al Aluminium 27, 14 Si Silicon 28, 15 P Phosphorus 31, 16 S Sulphur 32, 17 Cl Chlorine 35·5, 18 Ar Argon 40

Row 4: 19 K Potassium 39, 20 Ca Calcium 40, 21 Sc Scandium, 22 Ti Titanium, 23 V Vanadium, 24 Cr Chromium 52, 25 Mn Manganese 55, 26 Fe Iron 56, 27 Co Cobalt 59, 28 Ni Nickel 59, 29 Cu Copper 63·5, 30 Zn Zinc 65, 31 Ga Gallium, 32 Ge Germanium, 33 As Arsenic 75, 34 Se Selenium, 35 Br Bromine 80, 36 Kr Krypton 84

Row 5: 37 Rb Rubidium, 38 Sr Strontium 88, 39 Y Yttrium, 40 Zr Zirconium, 41 Nb Niobium, 42 Mo Molybdenum, 43 Tc Technetium, 44 Ru Ruthenium, 45 Rh Rhodium, 46 Pd Palladium, 47 Ag Silver 108, 48 Cd Cadmium 112, 49 In Indium, 50 Sn Tin 119, 51 Sb Antimony 122, 52 Te Tellurium, 53 I Iodine 127, 54 Xe Xenon 131

Row 6: 55 Cs Caesium, 56 Ba Barium 137, 57 La Lanthanum, 72 Hf Hafnium, 73 Ta Tantalum, 74 W Tungsten, 75 Re Rhenium, 76 Os Osmium, 77 Ir Iridium, 78 Pt Platinum 195, 79 Au Gold 197, 80 Hg Mercury 201, 81 Tl Thallium, 82 Pb Lead 207, 83 Bi Bismuth 209, 84 Po Polonium, 85 At Astatine, 86 Rn Radon

Row 7: 87 Fr Francium, 88 Ra Radium, 89 Ac Actinium

* 58–71 Lanthanum series
* 90–103 Actinium series

288